*Science, Technology and Medicine in Modern History*

General Editor: **John V. Pickstone**, Centre for the History of Science, Technology and Medicine, University of Manchester, England (www.man.ac.uk/CHSTM)

One purpose of historical writing is to illuminate the present. At the start of the third millennium, science, technology and medicine are enormously important, yet their development is little studied.

The reasons for this failure are as obvious as they are regrettable. Education in many countries, not least in Britain, draws deep divisions between the sciences and the humanities. Men and women who have been trained in science have too often been trained away from history, or from any sustained reflection on how societies work. Those educated in historical or social studies have usually learned so little of science that they remain thereafter suspicious, overawed or both.

Such a diagnosis is by no means novel, nor is it particularly original to suggest that good historical studies of science may be peculiarly important for understanding our present. Indeed this series could be seen as extending research undertaken over the last half-century. But much of that work has treated science, technology and medicine separately; this series aims to draw them together, partly because the three activities have become ever more intertwined. This breadth of focus and the stress on the relationships of knowledge and practice are particularly appropriate in a series which will concentrate on modern history and on industrial societies. Furthermore, while much of the existing historical scholarship is on American topics, this series aims to be international, encouraging studies on European material. The intention is to present science, technology and medicine as aspects of modern culture, analysing their economic, social and political aspects, but not neglecting the expert content which tends to distance them from other aspects of history. The books will investigate the uses and consequences of technical knowledge, and how it was shaped within particular economic, social and political structures.

Such analyses should contribute to discussions of present dilemmas and to assessments of policy. 'Science' no longer appears to us as a triumphant agent of Enlightenment, breaking the shackles of tradition, enabling command over nature. But neither is it to be seen as merely oppressive and dangerous. Judgement requires information and careful analysis, just as intelligent policy-making requires a community of discourse between men and women trained in technical specialities and those who are not.

This series is intended to supply analysis and to stimulate debate. Opinions will vary between authors; we claim only that the books are based on searching historical study of topics which are important, not least because they cut across conventional academic boundaries. They should appeal not just to historians, nor just to scientists, engineers and doctors, but to all who share the view that science, technology and medicine are far too important to be left out of history.

*Titles include:*

Roberta E. Bivins
ACUPUNCTURE, EXPERTISE AND CROSS-CULTURAL MEDICINE

Roger Cooter
SURGERY AND SOCIETY IN PEACE AND WAR
Orthopaedics and the Organization of Modern Medicine, 1880–1948

David Edgerton
ENGLAND AND THE AEROPLANE
An Essay on a Militant and Technological Nation

Jean-Paul Gaudillière and Ilana Löwy (*editors*)
THE INVISIBLE INDUSTRIALIST
Manufacture and the Construction of Scientific Knowledge

Thomas Schlich
SURGERY, SCIENCE AND INDUSTRY
A Revolution in Fracture Care, 1950s–1990s

Eve Seguin (*editor*)
INFECTIOUS PROCESSES
Knowledge, Discourse and the Politics of Prions

Crosbie Smith and Jon Agar (*editors*)
MAKING SPACE FOR SCIENCE
Territorial Themes in the Shaping of Knowledge

Stephanie J. Snow
OPERATIONS WITHOUT PAIN
The Practice and Science of Anaesthesia in Victorian Britain

**Science, Technology and Medicine in Modern History**
**Series Standing Order ISBN 0–333–71492-X hardcover**
**Series Standing Order ISBN 0–333–80340-X paperback**
(*outside North America only*)

You can receive future titles in this series as they are published by placing a standing order. Please contact your bookseller or, in case of difficulty, write to us at the address below with your name and address, the title of the series and one of the ISBNs quoted above.

Customer Services Department, Macmillan Distribution Ltd, Houndmills, Basingstoke, Hampshire RG21 6XS, England

# Operations Without Pain

## The Practice and Science of Anaesthesia in Victorian Britain

Stephanie J. Snow

palgrave
macmillan

First published 2006 by
PALGRAVE MACMILLAN
Houndmills, Basingstoke, Hampshire RG21 6XS and
175 Fifth Avenue, New York, N.Y. 10010
Companies and representatives throughout the world

PALGRAVE MACMILLAN is the global academic imprint of the Palgrave Macmillan division of St. Martin's Press, LLC and of Palgrave Macmillan Ltd. Macmillan® is a registered trademark in the United States, United Kingdom and other countries. Palgrave is a registered trademark in the European Union and other countries.

ISBN-13: 978–1–4039–3445–1 hardback
ISBN-10: 1–4039–3445–2 hardback

This book is printed on paper suitable for recycling and made from fully managed and sustained forest sources.

A catalogue record for this book is available from the British Library.

Library of Congress Cataloging-in-Publication Data
Snow, Stephanie.
Operations without pain : the practice and science of anaesthesia in Victorian Britain / Stephanie Snow.
p. cm. — (Science, technology, and medicine in modern history)
Includes bibliographical references and index.
ISBN 1–4039–3445–2 (cloth)
1. Anesthesia—Great Britain—History—19th century.
2. Anesthesiology—Great Britain—History—19th century. I. Title.
II. Series.
RD80.5.G7S56 2006
617.9'0941—dc22 2005044511

10 9 8 7 6 5 4 3 2 1
15 14 13 12 11 10 09 08 07 06

Printed and bound in Great Britain by
Antony Rowe Ltd, Chippenham and Eastbourne

*For my family, especially Evie, Verity and Gwyn*

# Contents

# List of Illustrations

# List of Abbreviations

| | |
|---|---|
| *AMJ* | *Association Medical Journal* |
| *BMJ* | *British Medical Journal* |
| *CB* | Ellis, Richard H (ed.), *The Casebooks of Dr John Snow* (London: Wellcome Institute for the History of Medicine, 1994). |
| *L* | *Lancet* |
| *LMG* | *London Medical Gazette* |
| *MT* | *Medical Times* |
| *MTG* | *Medical Times and Gazette* |
| *OC* | Snow, John, *On Chloroform and Other Anaesthetics* (London: Churchill, 1858). |
| *OE* | Snow, John, *On the Inhalation of the Vapour of Ether in Surgical Operations: Containing a Description of the Various Stages of Etherisation and a Statement of the Results of Nearly Eighty Operations in Which Ether Has Been Employed* (London: Churchill, 1847). |
| *ON* | Snow, John, 'On narcotism by the inhalation of vapours', *London Medical Gazette*, 1848–51. |

# Acknowledgements

Humphry Davy's words from his work on nitrous oxide encapsulate my own thoughts as the process of writing this book draws to a close:

> In the course of this investigation, I have met with many difficulties; some arising from the novel and obscure nature of the subject, and others from a want of coincidence in the observations of different [historians] on the properties and mode of production of [anaesthesia]. By extending my researches . . . and by multiplying the comparisons of facts, I have succeeded in removing the greater number of those difficulties, and have been enabled to give a tolerably clear history of the [innovation of anaesthesia]. . . . I have endeavoured to guard against sources of error; but I cannot flatter myself that I have altogether avoided them. . . . I cannot close . . . without acknowledging my obligations.

Indeed, this book has only been made possible through the help and support of numerous individuals. The project, which was generously supported by a Research Fellowship from the Wellcome Trust for the History of Medicine, was undertaken in the Centre for the History of Science, Technology and Medicine at the University of Manchester. The Centre's exemplary knack of combining intellectual rigour and friendliness makes it a pleasure to work in. Many archivists and librarians have provided me with invaluable help over the years and I would particularly like to thank staff of the Wellcome Institute for the History of Medicine, the Royal College of Surgeons of England, the British Medical Association, the Royal Society of Medicine, St George's Hospital, St Bartholomew's Hospital, King's College Hospital, the London Hospital, the Royal Free Hospital, the London Metropolitan Archives, Westminster Medical Society, Edinburgh Public Record Office, Glaxo Wellcome, the History of Anaesthesia Society, the John Rylands University Library and Liverpool Medical Institute. I thank Her Majesty Queen Elizabeth II for permission to cite correspondence between Queen Victoria and the Duchess of Sutherland, held by the Royal Archives at Windsor; Geoffrey Davenport, librarian of the Royal College of Physicians, who, many years ago, kindly supplied me with a copy of John Snow's casebooks; Robert Galland for sharing his data on amputations at the Royal Berkshire

Hospital; and Andrew Smith for giving me a copy of his draft report *Exploring Professional Boundaries in Anaesthesia*. For their kind permission to reproduce images, I thank the Director and University Librarian of the John Rylands University Library, the University of Manchester, St Bartholomew's Hospital Archives and Museum, the Wellcome Library, London and Michael Bennett-Levy of Early Technology.

For his generous choreography of ideas and arguments, as well as his persistence in asking the unanswerable, I owe much to John Pickstone. His comments on the entire manuscript, particularly his input to the first chapter, have served to make it a better and stronger work. I am also indebted to my colleagues who have taken the time to read various parts of this work: Ian Burney, Janet Greenlees, Aya Homei, Elizabeth Toon, Mick Worboys, and to the anonymous reviewer whose enthusiastic and useful report helped shape parts of the book. Other colleagues over the years have contributed through many ad hoc conversations, as have the students I have taught at CHSTM and the Open University, and delegates at the conferences of the British Association of Victorian Studies and the Society for the Social History of Medicine, where I have presented parts of my work. I also owe a debt of gratitude to the History Department of Keele University where, a long time ago, the seed for this work was planted. John Briggs and David Vincent were unfailingly supportive (and patient!) during the long emergence of my doctoral dissertation on John Snow.

But it is at the heart of my family where my greatest debts lie: my parents who have provided unstinting support, food and childcare, particularly during the writing period; Evie for loving me throughout and for helping with the 'little ones'; the little ones themselves – Verity and Gwyn – for whom this book has been both a source of pride and exasperation; and my husband, whose faith and love are my constants.

Stephanie J. Snow
Erway Hall
July 2005

# Introduction

On 1 December 1846, William Cowen, a 23-year-old stout and healthy-looking groom was admitted into St George's Hospital, London, having been thrown with great violence from a horse. His right thigh had been lacerated; the wound penetrated the muscles and exposed the bone, leaving the nerves and arteries almost bare. As December passed and the new year of 1847 began, inflammation and infection spread through Cowen's limb; by the end of January his knee-joint was distended with foul-smelling pus and it was feared that even a physical examination might hasten his death. The surgeons of St George's – William Cutler, Caesar Hawkins and Thomas Tatum – determined that Cowen's only chance of recovery lay in amputation and agreed that it should be performed using a new method of pain-relief.

Whilst Cowen had been lying in his hospital bed, news of a novel American technique of creating insensibility to surgical pain by administering ether vapour prior to operating had reached London. London's elite doctors and dentists had been captivated by the innovation and John Snow, a general practitioner in London's Soho area, was amongst the first to witness its use during a dental extraction. Snow soon began to give ether at St George's and on 28 January – his first session – he joined Cutler and his assistants at Cowen's bedside. Snow poured the pungent liquid into an inhaler, placed a mask over Cowen's mouth and held his nostrils in order to encourage his inhalation. In less than three minutes, Cowen became perfectly insensible. His body was drawn down to the foot of the bed and Cutler amputated the gangrenous limb. Whilst Cutler was completing the last stages of the stump repair, Cowen regained consciousness, saying he had felt nothing of the operation. By next morning, the success of the operation was apparent: Cowen had enjoyed an excellent night's rest, free of the usual 'startings' of the stump;

he even 'smiled when addressed' by the surgeons. Twelve days later Cowen was fit to leave the bed he had occupied for the previous 2 months; he gradually regained his health and was discharged on 12 May.[1]

Since 1846, anaesthesia has been the most powerful example of medicine's capacity to transform human experiences of suffering and pain. Those like Snow, who first saw patients like Cowen lie in 'perfect quietude' under ether and wake from surgery with no recollection of pain, recognised the innovation as epochal.[2] From the viewpoint of the twenty-first century, the status of anaesthesia within the history of medicine has not diminished. It stands as a defining moment; an irreversible divide between modern medicine and earlier practices. Stories like Cowen's make it easy to see why this is so; without ether, surgeons may well have decided the shock and pain of an amputation was too great a risk to undertake – he could have eventually died from sepsis. Thus from its earliest use, the innovation of ether succeeded in revolutionising the experience of surgery. The expectation that surgery will be painless remains indeed at the bedrock of our twenty-first-century medicine. It is hardly possible to construct the innovation of anaesthesia as anything but beneficial.

For this reason, the historical treatment of anaesthesia has been disposed towards the celebratory. Most histories of medicine chronicle the well-known events: the nitrous oxide experiments of Humphry Davy in the 1790s; the claims of the 'discoverers' of anaesthesia – Horace Wells, Charles Jackson and William Morton; the first use of ether vapour in Boston and London in 1846; and Queen Victoria's use of chloroform in childbirth in the 1850s.[3] Many full-length studies have been published and no group has been more vigorous in charting the history of anaesthesia than its practitioners themselves. Within this genre, Norman Bergman's recent account of the genesis of anaesthesia[4] follows the work of earlier anaesthetists such as F. F. Cartwright, K. Bryn Thomas, Thomas Keys, W. D. A. Smith and W. S. Sykes.[5] There are thriving anaesthesia history societies in Britain and the USA, anaesthetic textbooks carry potted histories of the specialty and specialist journals incorporate historical articles. All this suggests that the history of anaesthesia has resonance for twenty-first-century practitioners.[6]

But for all the work on the pioneers and on technical developments,[7] there are few histories which take a broader perspective, and some of the most historically revealing and fruitful lines of enquiry have been left untouched. Few historians, for example, have sought to explain why anaesthesia emerged in the 1840s, rather than an earlier period, though the gases used had been known for some time and inhaling vapours was an established therapeutic practice. When anaesthesia did emerge, it was from a serendipitous partnership of American gas sniffing and commercial dentistry, rather than from the vanguard of European medical science. Anaesthesia was rapidly taken up, but it was contested and in some ways strangely paradoxical. Inhaling gas brought patients to the verges of death to save them the agony of surgery. In the operating theatre, surgeons were accustomed to terrified or restless patients, but not to the newly insensible body which still breathed and might struggle in a way that inhibited surgery.

Most writers portray anaesthesia as a natural and inevitable phenomenon of 'modern' medicine which enjoyed an immediate and sustained take-up and developed uniformly. But this was not so: the early use was uncertain and differed between nations. Doctors and patients keenly supported the introduction of ether but it was a difficult gas to administer successfully and often caused excitement rather than insensibility. After the initial wave of experimentation many British doctors abandoned it. Anaesthesia may well have remained on the margins of British practice had it not been for the introduction of chloroform in 1847. Doctors found chloroform produced insensibility easily and effectively but the fatalities that followed its use established it as a 'high risk' anaesthetic. From the late 1840s medical communities in the northern American states and some parts of Europe chose to abandon chloroform and return to ether, but Britain, most of Europe and the southern American states continued to use chloroform. These differences reveal the way in which anaesthesia was shaped by the social and cultural expectations of medicine in each community: in London, surgeons were prepared to accommodate the risks of chloroform in exchange for its efficacy; in Boston, surgeons returned to ether, for fear that, in the event of a fatality, they would be sued for using a high-risk agent. And closer to home, although chloroform was used throughout Scotland and England, different methods of administration prevailed and doctors disagreed over the mode of chloroform death; on whether it killed through the respiration or the heart.

To skate over such complexities is to prevent the history of anaesthesia from informing our deeper understandings of nineteenth-century medicine and society. Pernick's work on ether's introduction to east coast America and its association with issues of medical professionalisation stands as a rare example of the insights which can be gained from putting anaesthesia under a wider historical lens.[8] A few recent historians have built on his work,[9] but still, 'much of the humdrum world of historical anaesthesia remains a vast undiscovered country'.[10] This book will address this deficit by examining complex patterns of innovation, reversals, debate and geographical difference by which anaesthesia became established in British medicine. In no way is it intended to be a linear history of discoveries, techniques or famous men. All these aspects are already covered in existing material and do not add to the more nuanced analysis I am seeking.

The book covers the period between 1790 and 1900, by which date the practice of anaesthesia was embedded into social and medical culture. Techniques such as regional, local and spinal anaesthesia and new anaesthetic drugs continued to evolve after this point, but all of this development was based on the conviction that it was socially and medically justifiable and necessary to remove the pain of surgery, childbirth and dentistry. I show how the establishment of anaesthesia as routine practice was achieved in two distinct phases. The first period stretching from 1846 to around 1860 was a time during which a question mark hovered over its viability – were its risks greater than its benefits? That issue is the heart of this book; it continued to be debated whilst patients and doctors grappled with the realities of painless surgery, the exhilaration of ether and the efficacy of chloroform – not just to remove pain, but to kill. But by the 1860s there had been a clear shift, driven in large part by patients' support of the innovation and anaesthesia was established as an integral part of the surgical experience. Yet the continuing incidence of anaesthetic fatalities meant that its practice remained firmly under public and medical scrutiny until well into the twentieth century.

I seek to illuminate three areas of current historical interest: the relationship between medical practice and science, the dynamics which structured patient–doctor relations, and the specialisation of medical practice. It is important to acknowledge at the outset that I have approached the first two issues from a particular historiographical vantage, using a typological model of medicine. Historians agree that there was a significant shift in both the practice and the intellectual construction of Western medicine around the end of the eighteenth century. Alongside the traditional concern with disturbances of the whole body, as described by patients, grew a concern with local lesions of body tissues, investigated by clinical examination and post-mortems. This new type of medicine is associated with the 'birth of the clinic', especially in post-Revolutionary Paris; and more generally with the claims that medicine could be a universal science of bodies, to which the social and geographical location of the patient was marginal. I suggest in Chapter 1, that it was indeed the new model of the body that emerged from this reconfiguration, which made explicable the artificial creation of insensibility using ether.

Throughout this work, I will draw on this model of medical types which has been successively developed by Ackerknecht, Jewson and Pickstone.[11] I will usually distinguish here between biographical medicine and scientific medicine, understanding the former to be compounded of the natural history of the body and disease, together with the symbolic and social meanings they held for patients and others. I shall use the term 'scientific medicine' to refer to the analysis of the body as it developed in the early nineteenth century, and also include the experiments, sometimes on animals, that were made by Snow and others. Sometimes I want to dissect each type,[12] but usually the simple opposition will serve my purposes, as indeed it structured the arguments of the protagonists. With Pickstone, I stress that scientific medicine did not, and could not, replace biographical medicine; rather the types structured continued debate about approaches and priorities.

Each type of medicine can be characterised by a particular knowledge structure and set of social relations. In biographical medicine the body is constructed as an individual and holistic system. Historians have frequently described the client–patron relationship which existed between wealthy patients and elite physicians in the eighteenth century as the epitome of this type. Scientific medicine, and especially analytical medicine, is usually typified by the model of the body found in the post-Revolutionary hospitals of Paris – that of a universal system of tissues and organs, largely discovered in the dead-rooms of hospitals. Chapter 1 will explore the differences between these two types of medicine in greater detail. The framework makes particularly meaningful two of my key interests: the relationship between the science and practice of anaesthesia, and the dynamics of patient–doctor encounters.

First, the interplay between the new scientific medicine and medical practice. One argument has been that, certainly up until the 1860s and the emergence of the laboratories, medical science was a rhetorical device, rather than a key influence on practice.[13] But I have previously shown, using the example of John Snow, that from the 1840s there were medical practitioners who did determine their practice through the use of chemistry and physiology.[14] Here I want to extend these arguments by exploring the relationship between analysis, experimentation and the practice of anaesthesia. Snow is used throughout this work as a key exemplar. He was committed to an ideology which wedded science to practice. From the outset, he embraced the theory of anaesthesia and pioneered his career as an anaesthetist in the metropolis. At the time of his early death in 1858, he was recognised nationally as London's first specialist anaesthetist. As well as two books on anaesthesia and many published papers, he left a set of casebooks which document almost 4500 anaesthetic administrations.[15] I will pull together data from these casebooks, his publications and the archives of London hospitals to tease out the relationship between anaesthetic theory and practice, and to show how many of our current historical assumptions about anaesthesia need re-evaluation.

In Snow's view, anaesthesia was a form of medical science; the findings of chemistry and physiology could shape clinical practice. But he also treated many private patients, whose elite status and sense of control over their medical destiny created an expectation that they should be treated as individuals, with due attention to *their* understandings. How did Snow mesh (or not) the cognitive processes of his practice of anaesthesia on patients who approached the encounter with these different expectations? Although this work in no way lays claim to being a 'history from below', it does seek to give shape to the vital part played by patient response to the innovation of anaesthesia, through the voices hidden in places such as surgical records and Snow's casebooks.[16] Analysing patient experiences of anaesthesia in hospital and private practice through the framework of these different types of medicine reveals stark differences in the access to anaesthesia during the early years of its introduction, primarily because hospital and private practice were characterised by different types of social relations.

Placing the innovation of anaesthesia at an intersection of these two different types of medicine has also helped contextualise the consistent presence of conflict within mid-nineteenth-century medicine. Journals of the period show clearly how medical practice was rife with disputes and disagreements over every aspect of theory and practice, as well as medical structures. And indeed, much work has been done on the sociological dynamics of this period.[17] There have, however, been few attempts to analyse the tension that doctors such as Snow perceived to originate from the juxtaposition of the 'new' scientific medicine against the 'old' biographical medicine of earlier generations. Using Pickstone's types as benchmarks, I can situate the response of doctors and patients to anaesthesia within a framework that illuminates and explains contradictory attitudes. So for example, I show that although Snow and the Scottish physician James Simpson were anaesthesia's strongest advocates, their very different approaches to its practice – scientific and biographical – stemmed from a fundamental divergence in their constructions of medicine and their own medical identities.

My third theme concerns the meanings of medical specialisation. Although there has been much work on the structure and workings of early and mid-nineteenth-century medicine, the emergence of medical specialties has been given far less attention.[18] There are few studies of individual specialties[19], and historians have commonly construed the process of specialisation within medicine as a manifestation of professionalism, or explained it as a sociological phenomenon – a means of achieving differentiation in an overcrowded marketplace.[20] Certainly each of these approaches feeds into my reconstruction of the dynamics of the specialisation of anaesthesia, but that emergence can also tell us much about the way in which the relationship between medicine and the public came to be built upon the notion of trust. This was not unique to anaesthesia. Throughout medicine, from the late eighteenth century, the knowledge-base of practice became closed to lay understandings. As patients became increasingly removed from an understanding of its processes, notions of trust rather than shared knowledge assumed an ever greater significance in patient–doctor relationships.

Anaesthesia is a particularly interesting example of medical specialism for it was a new practice without antecedent and had no direct heritage upon which to build. Other specialties grew out of the wider medical pool of practice. Rectal surgery, for example, emerged in the 1830s and specialists like Frederick Salmon employed general surgical principles – specialised anatomical knowledge and technical skills – to construct the theory and practice of a specialism which was carried out in a designated institution, St Mark's Hospital.[21] Anaesthesia could not be characterised in the same manner as other specialties – by the patient's disease, body part or stage of life. All surgical patients, from the newborn to the elderly, were potential recipients as were dental patients and mothers in childbirth. Nor too can its evolvement be linked to a specific medical, social or political event as with the burgeoning of orthopaedics during the First World War,[22] or the influence of the epidemic of Egyptian ophthalmia upon the beginnings of ophthalmology in the 1800s.[23] Remarkably, its creation as a distinct specialist practice during the last decades of the nineteenth century was unique to English, and particularly London practice. Here, anaesthesia did not develop simply as an adjunct to the surgical process, as say the new antisepsis practices of the 1870s. Instead, aspiring specialists drew together a set of physiological processes and practices involving particular gases and techniques to form a body of knowledge and practice that became an independent authority to surgery and was firmly placed in the medical domain. By the 1890s, London hospitals and some of the larger provincial institutions had appointed anaesthetists to their staff. In Scotland, America and most parts of Europe, however, anaesthesia remained under the surgeon's control and was a duty delegated to a junior member of the team, or to a nurse. Nor was the impact of these international differences short-lived. We shall see in the final chapter how today's anaesthetists face dilemmas that have their origins in this historical divergence. The emergence of specialist anaesthesia therefore tells us much about the influences of local context upon medical practice and shows how historical enquiry can illuminate the challenges of our times.

There are two important aspects of anaesthesia which I have had neither the time nor the space to give proper attention to here: dental and military anaesthesia. From the first, anaesthesia was perceived as an innovation that had particular resonance for dentistry and dental patients were some of the first to experience ether. Some work has been done on the development of eighteenth-century dentistry, but little on nineteenth-century practice, particularly its take-up of anaesthesia and the way in which this provoked a strong conflict between medicine and dentistry over issues of professionalisation.[24] By the late 1850s, anaesthesia had become a pivotal part of elite dental practice. Many of the dentists noted in Snow's casebooks held hospital appointments as surgeon-dentists and several of them were key proponents of dental reform, including John Tomes.[25] Dental anaesthesia formed a substantial part of Snow's practice and in Chapter 5, I sketch out the particular trends suggested by his records, but there is much left to say on the subject. So too, could we benefit from further work on the use of anaesthesia in the military context. In many respects, the battlefield produced different imperatives to those found in metropolitan practice, but the concern about the 'depressive' nature of chloroform and the stimulant of the 'smart of the knife' provoked a debate which straddled both communities.[26] The use of anaesthesia on the battlefields of the Crimean and the Franco-Prussian wars of the mid-nineteenth century has received some attention but much useful comparative work could be done on the differences between the use of anaesthesia in these European military contexts, and its employment in the American Civil War.[27]

For purposes of clarity, I use the term 'anaesthesia' throughout to refer to the inhalation of narcotic vapours. The proliferation of terms used to describe the process of anaesthesia – ethereal insensibility, etherisation, suspended animation, chloroformisation – testify to the innovative status of the technique. Anaesthesia soon became the most common descriptive, although 'chloroformisation' remained popular for much of the nineteenth century.[28]

## Overview

In Chapter 1, I set the scene for the introduction of ether by suggesting that the form of its emergence depended upon a key shift in the configurations of medicine which occurred between the 1790s and the 1840s. My analysis of the work of Davy and Hickman suggests that a new concept of an unfeeling, yet living, body became possible through the new physiological and anatomical knowledge of the 1820s onwards. Thus by the 1840s, there was an elite group within the London medical community who were receptive towards the idea of the artificial suspension of sensibility within a body, without compromising respiratory or circulatory functions. When news of the Yankee dodge with ether arrived in London, this elite group was well-placed to ratify its effects and place it in existing chemical and physiological frameworks. In Chapter 2, I describe these early trials of ether and chloroform and the key role played by Snow in the establishment of a framework for its practice. The introduction of anaesthesia threw into sharp relief, the differences between those doctors like Snow who perceived anaesthesia as a process which acted universally on all bodies, and those who, drawing upon their understanding of bodies as individual, holistic systems, perceived it to be unpredictable and fraught with danger. Snow and Simpson were strong advocates of the concept of painfree surgery, childbirth and dentistry; yet their methods of practice were very different and Chapter 3 shows how each man's technique was revealing of key differences in the concept of bodies and medical identities. Chapter 4 examines the way in which the risks of the process were quantified and ratified across medicine and society, and considers the paradoxical manner in which anaesthesia introduced a new fear into patient experience – the dread of unconsciousness. The first 15 years of anaesthetic practice in London are the focus of Chapter 5. Snow's anaesthetic practice data and information from the archives of London hospitals are used to examine the early spread and take-up of anaesthesia. I show how anaesthesia reconfigured the meanings and practice of surgery and explore the differentiation in access to anaesthesia experienced by hospital and private patients. Chapter 6 analyses the emergence of specialist anaesthesia in England between the 1860s and 1900 and shows how its structures – a professional society, journal and communities of practitioners – were constructed in the name of public safety. By 1900, anaesthesia was established as a routine process for all patients undergoing surgery and the concluding chapter considers what its history reveals about the complex relations between the practice of medicine, culture and society.

# 1
# From Enlightenment Philosophies to Victorian Reform, 1790–1846

The definitive beginning of anaesthesia is commonly taken as William Morton's use of ether in Boston during the autumn of 1846. From this point, there is historical consensus on the mode of its diffusion to Britain and Europe and on the alacrity of its worldwide adoption as an efficacious method of pain-relief. But anaesthesia has a curious pre-history, including the explorations of Humphry Davy, Henry Hill Hickman, Crawford Long and Horace Wells. There is agreement amongst historians that such figures are part of the history of anaesthesia but little resolution as to how their work can be integrated with later history; or explanation as to why anaesthesia did not emerge prior to 1846; or why exactly it was a dentist, rather than a surgeon, who succeeded in establishing ether.

Take for example, Bergman's recent account of the genesis of surgical anaesthesia. He describes in great detail the work of Davy and Hickman, yet does not grapple with the reasons as to why their experiments did not popularise anaesthesia.[1] Cartwright had previously analysed Hickman's experiments with 'suspended animation', but he failed to find a connection with later experimenters.[2] Smith concluded that Davy's researches set the stage for anaesthesia 'but the actors went away'.[3] One of the most recent pieces of writing on Davy, by Jacob and Sauter, offers a useful starting point from which to address these issues,[4] but in general the historical treatment of this prelude to anaesthesia, from the 1790s to the 1840s, has been deeply unsatisfying.

This chapter suggests that we can better 'place' the early work of Davy and Hickman *and* the later developments, by recognising how much the configurations of medicine had changed between the 1790s and the 1840s. I shall argue that from the later eighteenth century pain emerged as a key medical problem, partly through the new attention given to the experience of death. As patient tolerance of pain declined, the use of opiates grew and it was in this context that enterprising doctors were extending the range of operative surgery. This surgical project was increasingly linked to an anatomical view of the body, as a system of tissues and organs that supported each other to maintain life. When one part failed, or links were broken, the body died – but through a process which might be interrupted or even reversed, so life might be reclaimed. By the 1830s, physiologists and elite doctors envisaged a level of unconscious life, linked with the vegetative nervous system but divorced from the higher functions and the mind, including suffering. At the same time, the emergence of more conservative and slower surgery intensified the problem of pain for both patient and surgeon. By the mid-1840s pain no longer seemed physiologically necessary or socially acceptable; but the intensive use of drugs known to diminish surgical pain was dangerous, and new alternatives such as mesmerism were highly contentious.

I do not claim that these shifts in the practice and theory of medicine fully explain why ether anaesthesia entered medicine in the 1840s, but they certainly provide the practical and intellectual context within which that innovation was understood. By exploring the shifts in medicine over the previous century, we can reconstruct the world in which the 'Yankee dodge' of an unknown American dentist was accepted and developed by the medical elites of Britain and Europe.

To get a measure of the shifts, we can turn first to the celebrated experiments of the 1790s when Humphry Davy, later a very famous chemist, experimented with nitrous oxide, found it intoxicating and speculated about its possible use in surgery. For many commentators, this was an opportunity missed; for the historian it is a chance to explore how gases and the body were understood at the end of the Enlightenment, amongst England's most 'progressive' doctors, chemists and natural philosophers, just before the political and cultural reaction against the French Revolution. *For us*, Davy can exemplify the physiology of sensibility and enhancement as medical attention began to shift towards man's biological limitations and to seeing life through the window of death.

## Davy, gases and the enhancement of life c.1790

From the 1750s onwards, a series of discoveries had placed the nervous system at the centre of the body's capacity to experience both health and disease.[5] That stimulation of living bodies gave rise to sensations and motion was commonplace, but it was the experiments of Swiss physician, Albrecht von Haller, in 1752 which located sensibility as an exclusive physical property of the fibres of the nerves, and it was the work of Edinburgh physician, Robert Whytt, on the spinal cord and involuntary movement which grounded sensibility as a physiological function.[6] The emphasis upon the nervous system was reinforced by those such as William Cullen, medical doyen of Edinburgh University, who understood its qualities to be 'vital', visible and beyond mechanical explanation.[7] The immaterial soul furnished the nervous power or energy which drove the material body, and thus the body and soul were unified by sensibility. Heightened sensibility was perceived as the means of the 'perfectibility' of individuals and societies – a means of realising Enlightenment aspirations for progress.[8] This new physiology drew upon John Locke's philosophy of the mind – that all individuals were born with a mind which stored and shaped the data acquired through the senses, thus creating empirical knowledge of the world, which in turn gave individuals the power to transform their environment.[9] Within Enlightenment culture this monistic understanding of the person forged connections between sensibility and the aspirational values of 'polite' society; consumerism took hold, not least in the form of health and well-being, and health emerged as a new commodity.[10]

Bodies – constituted from solid fibres and numerous fluids – were understood by both patients and doctors to possess an individual and natural equilibrium. Disease or sickness caused an imbalance or disharmony to this equilibrium. The role of the doctor was to listen to the patient's narrative of his/her illness and observe the symptoms.[11] It was a relationship in which both participants accorded pivotal importance to the individual experience of illness and spoke the same language of sickness and disease. The weighting that was given to the whole body, rather than its parts, was reflected in the status accorded to physicians and surgeons. Physicians, by virtue of their university education, were deemed to be custodians of an elite knowledge which they used to treat elite patients.[12] Surgeons with their craft origins treated parts of the body – they let blood, drained abscesses, cut off tumours, pulled teeth and trussed ruptures. But the major operations of lithotomy or amputations were rare because the risks of haemorrhage or sepsis were high.[13]

Enthused by the prospect of discovering medical laws like those of Newton's in physics, doctors sought unitary explanations of disease that mirrored the 'idea of perfectibility' articulated in many Enlightenment writings.[14] The new stress on the nervous system was reinforced by Cullen's claim that 'almost the whole of diseases of the human body might be called NERVOUS'.[15] His work was extended by several of his students, particularly John Brown and Benjamin Rush.[16] Brown explained sickness in the body as an imbalance of irritability, which he described as excitability; in his 'Brunonian' medicine, sickness and disease were the consequences of this disequilibrium, and harmony could be restored in the individual using a straightforward set of therapeutic principles.[17] These were exemplified in his thermometer, which used temperature to ascertain the exact degree of under- or over-stimulation of the body, and thus the dose of therapy required.[18] Depressants such as bloodletting, emetics or cathartics reduced excessive excitability and stimulants such as opium or alcohol cured those whose excitability was depleted.[19] Life and health were inextricably bound with the external environment of the body. Brunonian medicine became far more popular on the Continent, particularly in Italy and Germany, than in Britain.[20] But it is of particular significance for our discussion because it was the theory which underpinned Thomas Beddoes and Davy's work on the therapeutic use of gases.[21]

Beddoes was very much a man of his time.[22] Trained in Edinburgh, he understood substances like airs or metals to be simple or compound by virtue of their 'elemental' qualities which resisted decomposition.[23] Chemistry was perceived as the epitome of Enlightenment values: 'I regard every experiment that Priestley made in chemistry as giving *wings* to his more sublime theological works', wrote his friend Samuel Taylor Coleridge.[24] Beddoes' support of the philosophies of the French Revolution – liberty, the rights of man, freedom from state oppression – forced him to leave a highly promising position as chemical lecturer and experimenter at Oxford in 1793 and to set up practice in Bristol. He took the opportunity to put into practice his dreams that chemistry could transform medicine.

Beddoes had been inspired by the chemical revolution, spearheaded by the French chemist Antoine Lavoisier, which isolated nitrogen, hydrogen and oxygen and brought about a new understanding of the 'different kinds of air' which made up the atmosphere. His intention was to use these newly discovered respirable gases to develop therapies for lung conditions such as tuberculosis. His connections to the Edgeworths, the Wedgewoods and the Watts through the Lunar Society provided support, funding and apparatus and in 1799 he opened the Pneumatic Institute in Clifton, Bristol.[25] The city housed some of the most outspoken radicals of the period and during its short existence the Institute became a magnet for enlightened thinkers like Coleridge, Robert Southey and Davies Giddy – later President of the Royal Society. Chemistry, philosophy and poetry were soulmates within its walls.

Davy's nitrous oxide experiments formed part of Beddoes' wider enquiry into the therapeutic use of gases. Beddoes' recruitment of the brilliant young chemist has been well-told,[26] but what needs to be stressed is the way in which Beddoes and Davy's radical understandings of natural philosophy gelled as strongly as their belief that chemistry was the means of discovering the powers and forces of life. Prior to joining Beddoes, Davy had undertaken experimental work on the nature of light and heat and concluded that the oxygen breathed by living beings was composed of oxygen gas and light – phoso-oxygen. Through respiration, the light was diffused through the nerves and provided the essential stimulation for the vitality of all bodies. For Davy it was a convincing explanation of materialism, a way in which 'one law alone may govern and act upon matter . . . the law of animation, tending to produce the greatest possible sum of perception, the greatest possible sum of happiness'.[27] And because Davy understood light to be a chemical, he claimed that chemistry 'in its connection with the laws of life' was 'the most sublime and important of all sciences', a view shared by Beddoes.[28]

At the Pneumatic Institute, Davy investigated the therapeutic potential of a range of gases – nitrogen, hydrogen, oxygen and hydrocarbonates. The first stage, using animals, was to ascertain if a particular gas was capable of sustaining life, and if so, did it stimulate or depress vitality? Through respiration, the gas was understood to enter the blood and travel to all parts of the body. It thus had the potential to rebalance, through stimulation or depression, the body's equilibrium. Davy's investigations showed that all gases apart from atmospheric air had the power to destroy life but in different ways.[29] Nitrogen and hydrogen gases, of themselves, caused no changes in the venous blood of animals; they died from a 'disease produced by privation of atmospheric air, analogous to that occasioned by their submersion in water'.[30] When he breathed hydrogen, Davy noted, it was 'the pain of suffocation' that compelled him to leave off the gas.[31] Oxygen, and gases made from compounds of hydrogen and carbonates, however, destroyed life by changing the blood in a way which made it 'incapable of supplying the nervous and muscular fibres with principles essential to sensibility and irritability'.[32] Nitrous oxide's mode of action was different; Davy found it could be respired for longer than any other gases except air and oxygen. Its effects on animals were to cause an initial period of excitement which was followed by exhaustion. If the animal stopped breathing gas before complete exhaustion was reached, then it was possible to restore 'healthy living action' by letting the animal respire atmospheric air.[33]

The 'peculiar changes' in the blood and organs of animals then were reversible – but Davy certainly understood them as a process of death, although different to that caused by the privation of air.[34]

The conception of death as a process, rather than an absolute, was very recent. The debate had begun in 1740 when Jacques-Benigne Winslow, Professor of Anatomy in Paris, had suggested that the state of death was potentially uncertain. An absence of pulse or a cessation of breathing were not irrevocable states. Only when the flesh of the body began to rot, claimed Winslow, was an irreversible state of death reached.[35] By 1792, James Curry in Liverpool noted that 'the happy discovery of an essential difference between absolute and apparent death' had lately changed the previously universal belief that 'life quitted the body in a very few minutes after the person had ceased to breathe'.[36] Death, viewed as a process, sustained the possibility of new medical roles and attention began to focus on resuscitation and the various techniques of hot/cold baths, galvanism and physical manoeuvres that might be used to restore an apparently dead body to life. In Amsterdam, a society had been founded in 1767 for the recovery of drowned persons and London followed this example in 1774, by establishing an institution for affording immediate relief to persons apparently dead from drowning.[37] It later became the Royal Humane Society.[38] Its objective was to teach resuscitation techniques, particularly in cases of water accidents, many of which occurred in the Serpentine Lake in Hyde Park as well as in the Thames. John Hunter, London's leading surgeon of the time proposed the use of two-chamber bellows to artificially ventilate the lungs; he also drew on the recent discovery of oxygen and its stimulant properties by suggesting that the gas could be administered rectally using a small pair of bellows.[39] Davy's suggestion that nitrous oxide could be used to restore life in drowned or suffocated bodies shows his absorption of these new understandings of death.[40]

And for these understandings, nitrous oxide had a peculiar importance. Joseph Black and Joseph Priestley had isolated nitrous oxide in the 1770s,[41] and Davy's attention had been drawn to the subject by the claim of Samuel Mitchill, professor of chemistry in New York, that the gas was produced in the body by the decomposition of food. In healthy bodies the gas was simply excreted but in illness it promoted fever and was lethal if inhaled.[42] Fever, and its links with confined spaces and rebreathing, was another major issue of the day. When Davy succeeded in isolating pure nitrous oxide, he used animals to test its action, and then, believing that it was respirable, proceeded to inhale the gas himself. And thus his attention was turned from disease and death towards the enhancement of life. After a couple of attempts he breathed enough to experience a 'highly pleasurable thrilling…the objects around me became dazzling and my hearing more acute…the thrilling increased, the sense of muscular power became greater'.[43] His immediate extension of these experiences to Beddoes and friends underlines the profound intellectual importance he attributed to the state of altered sensibility. For him it was a life-affirming process: 'vivid ideas passed rapidly through the mind', 'sublime emotions connected with highly vivid ideas', 'I existed in a world of newly connected and newly modified ideas; I theorised, I imagined I made discoveries', he noted during his trials.[44] Breathing nitrous oxide brought about physical changes in the body – muscular power, tingling skin – but most striking for Davy and others were the altered intellectual sensibilities. The sensations created by the gas were beyond existing experience. 'Davy has invented a new pleasure for which language has no name', gloried Robert Southey whose breathings had produced a vision of 'a paradise wholly immaterial – trees of light growing in a soil of ether – palaces of water refracting all with colours'.[45] And James Thomson complained that the English language was so 'defective' that to comprehend the effects of 'this extraordinary gas' one had to respire it and then 'either invent new terms to express these new and particular sensations, or attach new ideas to old ones'.[46] The way in which nitrous oxide intensified individual engagement with the world dovetailed neatly with Enlightenment aims of self and social improvement. It struck a chord with Brunonian theory that vitality depended on stimulation – and the greater the stimulation, the higher the state of perception.[47] It was, noted Davy, the 'intellectual pleasure, or hope' induced by the gas which made many so strongly inclined to breathe it again.[48]

During one of his trials of nitrous oxide, Davy was suffering from the 'intense physical pain' of toothache. He noted that it 'always diminished after the first four or five inspirations' of gas. As he breathed in, he experienced a 'thrilling' which 'swallowed up in pleasure' the unpleasant sensations of his pain.[49] When he later published his work, he concluded that the powers of nitrous oxide 'may probably be used with advantage during surgical operations in which no great effusion of blood takes place'.[50] His suggestion was not taken up in any of the contemporary reviews of his work and a satisfactory explanation of why Davy or his contemporaries did not pursue the possibility has so far eluded historians. I want to suggest that some light can be shed on the question by examining Davy's understandings of bodily sensibility and by linking them to the wider social and political context of the late 1790s.

For Davy, the nervous system was the very self of the individual body; it was supplied through the blood with the principles essential to sensibility and irritability. He explained the mechanism of nitrous oxide through his understandings of sensibility. 'Reasoning from common phenomena of sensation, particularly those relating to heat', he wrote, 'it is probable that pleasurable feeling is uniformly connected with a moderate increase of nervous action; and that this increase when carried to certain limits, produces mixed emotion or sublime pleasure; and beyond those limits occasions absolute pain'.[51] Davy supported the use of opiates and alcohol to ameliorate physical pain; nevertheless he understood pain to fulfil a physiological function as its presence marked the return of vitality to the body after illness. 'By whatever cause the exhaustion of organs is produced', he remarked, 'pain is almost uniformly connected with their returning health'.[52] Certainly Davy, Beddoes and their circle embraced the Enlightenment quest to diminish human suffering. They held fast to the hope that 'at some period' physiology would 'become a branch of philosophy... interested in teaching the means of procuring pleasure and removing pain'.[53] But in 1800, it was a means that they did not believe they possessed.

Nor too was the social context conducive of further explorations into nitrous oxide gas. During the 1790s, whilst Beddoes was establishing his institution, fear had grown amongst British conservatives that the anarchy of the French Revolution would spread across the channel.[54] A revulsion against all things French caused many areas of natural philosophy to become politicised;[55] gas chemistry was particularly susceptible, partly because Lavoisier was perceived as the pioneer of the 'new' chemistry and partly because of the materialistic implications of the subject.[56] The publication of Edmund Burke's *Reflections on the Revolution in France* in 1790 had marked the beginning of a scathing condemnation of Enlightenment philosophies by conservatives and placed particular emphasis upon the dangers of the 'new chemistry'.[57] Beddoes had not moderated his radical views since leaving Oxford; for him, medicine, society and politics were interdependent and the new gas chemistry was redolent of the 'perfectibility' of society. Burke argued that such claims were not only false, but most dangerously threatened the moral basis of humanity. Thus by the turn of the nineteenth century the Pneumatic Institute and the nitrous oxide experiments, in particular, were ripe for criticism and satire in conservative publications such as the *Anti Jacobin Review*.[58] Chemistry was not the only area of concern. Mathematics too was considered to be open to dangerous French influences. In 1801 John Robison, a conservative professor of physics at Edinburgh, wrote new articles for the 3rd edition of the *Encyclopaedia Britannica*; he specifically rejected French algebraic methods and spoke of the 'seeds of Anarchy and Atheism' which were spread through the French *Cyclopaedie*.[59] Animal magnetism and mesmerism were also sidelined because of their close associations with France.[60] By 1801, the Bristol experiment had come to an end; Davy had departed to the Royal Institution and Beddoes converted the Pneumatic Institute into a Preventive Institution 'for the benefit of the sick and drooping poor'.[61]

The Royal Institution to which Davy moved had been founded in 1799 as a philanthropic venture – the Society for Bettering the Condition of the Poor. Berman has shown how it was established by a group of landowners who were concerned that the intense rural poverty of the 1790s would create social unrest. The intention was to improve the social and working conditions of the rural poor through improved husbandry techniques, the building of cottages and soup kitchens.[62] When Davy joined the Institution in 1801 he took up research into tanning and agriculture, combining chemical investigation with analysis of current practices, and it is evident that his patrons expected his experiments to produce economic paybacks for the tanning industry.[63] The nature of the project was far more utilitarian than the gas research undertaken at the Pneumatic Institute, and Golinski has convincingly argued that Davy came to see himself as an expert instructing an audience.[64] It was worlds away from the collective experiences at the Pneumatic Institute.[65] Davy's lecture notes suggest that the shift was already clear by 1802:

> we do not look to distant ages, or amuse ourselves with brilliant though delusive dreams concerning the infinite improveability of man, the annihilation of labour, disease, and even death, but we reason by analogy from simple facts, we consider only a state of human progression arising out of its present condition.[66]

The British rebuttal of radical Enlightenment philosophies and the revitalisation of conservative values during the early 1800s can also be seen in literature, for example in Jane Austen's *Sense and Sensibility*. Written between 1797–98, although not published until 1809, the novel juxtaposes Elinor's 'coolness of judgement' and her ability to control her 'strong feelings' against Marianne's lack of moderation and excessive sensibility. Marianne's success in eventually taming and reordering her excessive sensibilities was signalled by a new social order in which she became matriarch of new domestic and social duties.[67] So too at the

Royal Institution, Davy succeeded in controlling and refocusing the excessive sensibility of the nitrous oxide experiments. Beddoes was not so fortunate. He died aged 48 years in 1808, depressed and possibly suffering the effects on heart and lungs of his self-experimentation.[68]

It is indisputable that Davy's work on nitrous oxide has a place in the history of anaesthesia; his was the unequivocal demonstration of the power of gases to alter bodily states. That neither Davy nor his contemporaries pursued the possibilities of nitrous oxide is understandable *if* we place his work within the context of his 1800s understanding of the body and its processes. For Davy, pain and pleasure were polarities of sensibility grounded in the nervous system. Nitrous oxide had the power to transform painful sensations to pleasurable ones by increasing the nervous energy within the body, but he did not conceive of a way in which sensibility could be disassociated from the body without adversely affecting its living principles. For these reasons, his suggestion that nitrous oxide might serve a purpose during surgical operations should be read as a means of using its stimulant qualities to counter the depressive and painful ones of surgery, as he had experienced during toothache and headache. Davy understood the suspension of sensibility to be a form of suffocation which ended in death, or at the very least, caused intense pain during its return.[69] Intervening into the body's processes to artificially suspend sensibility was a hypothesis that was physiologically unsupportable, despite his clear support for the humanitarian mission to remove pain at all levels of human experience.

## The control of pain

No one continued Davy's work on enhanced sensibility, but many continued the late eighteenth-century interest in opiates and pain reduction. Though several historians have suggested that researches before 1840 failed to establish anaesthesia because the amelioration of pain was not then a social objective, this argument cannot be sustained.[70] We can accept that the experience of pain is not a constant, that it is always shaped and defined by its cultural context, but we have good evidence that from Greek times onwards societies have variously sought to understand and ease the pain of chronic disease, surgery and childbirth.[71] Moreover, it is now clear that the last decades of the eighteenth century saw a burgeoning in the use of opiates. In the words of Porter, it would appear that the very pain threshold of society was becoming lowered.[72] And unlike the Bristol cavortings, opiates for the sick and dying chimed well with strengthening evangelical concerns over the manner of one's death.

For most of the eighteenth century, the priest rather than the doctor had held sway over the deathbed, reflecting the cultural dominance of the religious rituals of preparing the soul for death. But in parallel with the novel techniques of artificial respiration which had grown out of the new view of death as a process, doctors began to provide comfort to the dying.[73] Thomas Percival, physician at the Manchester Infirmary, called for doctors to 'smooth the bed of death' 'by alleviating pain and soothing mental anguish'.[74] Management of death became part of the definition of a 'good' doctor: from the 1800s, the London physician, Henry Halford, became 'a master in all that concerned the management of dying', not just through the use of pain-relief but through his sympathy and the hope he inspired.[75] This new medical role at the deathbed was bolstered by the philosophies of evangelicism which emerged as a decisive engine of social, political and medical change over the first decades of the nineteenth century.[76]

Because evangelical doctrine supported the possibility of individual salvation right up to the final moment of death, the final hours and days of life were charged with great significance; a rapid death was feared because it allowed no time for spiritual repentance. Publications such as the *Evangelical Magazine*, published between 1793 and 1892, dwelt on true experiences of deceased evangelical Christians.[77] If doctors were able to ameliorate or at least palliate physical pain, then the dying patient could remain lucid enough to make final repentances, and family and friends could mourn them in the knowledge that they had been truly saved. A lady's maid who died from a rupture of the stomach in 1838 was attended by Suffolk surgeon Edward Crowfoot who observed how the use of opium kept her 'perfectly sensible to the last, suffering but little pain'.[78] It certainly appears that one of the most valued qualities of opiates was their effect upon the mind; opium 'affords to the mind a peculiar energy, elevation and tranquility', noted Prussian physician, Christian Hufeland.[79]

The idea that pain could be avoided broke sharply from established understandings which believed that pain performed a vital function within the body's systems. Pain was an 'unpleasant sensation or irritation' noted *Hooper's Medical Dictionary* in 1820. It was a 'voice of nature', a protective device which could warn of internal inflammation or disease in advance of visible symptoms.[80] The physical discomfort created by many therapies – bleeding, emetics, purging – was understood as 'pain for gain' and, as Davy had noted, the return of vitality to the body was usually accompanied by unpleasant sensations.[81] Therapeutic practices reflected the functional nature of pain.

John Hunter's principles of healing, for example, were based on the understanding that diseases and wounds caused 'irritability' which disrupted the 'universal sympathy' within the body. By employing the principle of counter-irritation at a different site on the body, healing would be encouraged in the primary inflammation or irritation. For this reason, techniques such as blisters, cautery and acupuncture were used, either at the site of wounds after injury or operation, or on other parts of the body in cases of chronic disease such as neuralgia or rheumatism. Stanley details the acutely painful nature of many of these therapies but there is no question that surgeons promoted such techniques and patients bore them because they were understood to offer the best chance of recovery and return to health.[82]

But by the 1830s the radical view that pain was purposeless began to emerge. An 1838 treatise by the Dublin doctor, James Macartney, suggested that the healing of wounds could take place without any need for counter-irritation of the site. It was the 'most original medical work' since the time of John Hunter, pronounced the *Lancet*. Macartney understood his theory to reflect 'the humane spirit of the present age'. The practice of promoting inflammation of wounds came to us 'from those dark ages of the world', he noted:

> in which insanity was treated by the whip and chains, when people were forced to profess their belief or impossibilities by the rack or the faggot, when the punishment of death was awarded to almost every crime.[83]

New strategies of employing opiates to palliate the pain of childbirth began to be adopted. Brentford practitioner F. A. B. Bonney found them useful in the early stages of labour to quiet the 'tedious pains which appeared to be useless,' as well as employing them in the post-natal period. It was a notably different understanding to that of J. Paterson of Aberdeen, who claimed the 'after-pains' of labour should be encouraged because 'such pains are doing good somewhere'.[84]

Patients too grew to have different expectations of therapeutics. Witten has shown how German patients in the 1820s and 1830s were keenly receptive to the gentle homeopathic therapies of Samuel Hahnemann.[85] And Parssinen suggests that the increased use of opiates in the 1830s was an attempt by orthodox practitioners to compete with unorthodox practitioners who were offering patients less painful and more palatable remedies.[86] Opium was without doubt the jewel in the materia medica and by the 1840s was the mainstay of therapeutic practice.[87] Indeed, as Berridge and Edwards have noted, it is almost easier to list the conditions where its use was excluded, rather than vice versa.[88] The use of opiates therefore empowered doctors to effectively lessen the pain of chronic disease, the suffering of death and that of childbirth. What remained problematic and appeared increasingly so in the context of the growing complexity and range of surgical procedures was the pain of the knife. To understand that growth in surgery, we return again to the period of the French Revolution.

## Revolutions of the body

As Fissell and Jacyna have shown, Britain had its own surgical tradition, which blossomed from the later eighteenth century, especially in the new teaching hospitals and in the military.[89] Surgeons sought every opportunity to extend their skills through dissection or operations and such occasions drew crowds of other surgeons and apprentices who were equally intent on expanding their knowledge. But British hospitals were run as charities by lay governors who depended on subscriptions and who were wary of any suggestion that doctors were experimenting on patients, living or dead. And though surgeons were rapidly gaining status, they remained in some ways subordinate to physicians. Surgeons thought of the body as a set of organs and tissues that might be manipulated in various ways, but the reigning view of the patient belonged to the physicians: the body as a holistic system with an individual equilibrium. And in Britain, in medicine as in politics, change was gradual.

By contrast, the Revolution in France, by removing old institutions and setting up a new medical school, encouraged a bold new construction of the body which, during the first decades of the nineteenth century, was taken up by keen young doctors in Britain and America.[90] A new state-controlled medical system reformed the large hospitals in Paris, offered salaried posts to surgeons and physicians, supported clinical teaching on the wards and facilitated the dissection of corpses. Thus the surgical view of the body could be extended and built into the new curriculum. Pathology and anatomy became the key disciplines: post-mortems allowed doctors to locate tissue lesions – inflammation, tumours, sepsis – which could then be correlated to the symptoms observable in the sick. The large hospitals facilitated comparison and grouping of patients; the use of statistics – the numerical method – was introduced, and the relationship between patient and doctor was changed. Physical examination of the patient, using new methods of diagnosis such as the stethoscope, became the mainstay of the consultation, displacing the patient's account of their illness.[91] The objective was to classify the symptoms of each and every patient within a universal pathology and for this reason from the 1820s onwards the benchmark of 'normal' rather than 'natural' began to be applied to bodily phenomena such as the pulse, urine output or heartbeat.[92] The new approach derived heavily from surgery and it supported the extension of surgical practices. If many diseases, fundamentally, were local lesions of tissues, then in principle they might be susceptible to surgery.

In Britain, a new surgical elitism based on the expanding fields of anatomy and pathology was already evident by the early nineteenth century.[93] The College of Surgeons had separated from its craft links with the barbers in the mid-eighteenth century, and the work of the Hunter brothers – the physician, William, who established a medical school in Great Windmill Street and John, the surgeon whose 'principles' became constructed as the foundation of 'modern surgery' – epitomised the way in which pathology and anatomy emerged as the key disciplines.[94] The new emphasis on dissection created a particular problem in Britain; the only legitimate corpses were those of hanged murderers and thus the 1820s saw an intense conflict between the demand and the availability of subjects for anatomists in the medical schools. Richardson's study shows vividly the complex medical and social contest during these years over the availability of bodies, and how this led to the passing of the Anatomy Act of 1832.[95] The focus on anatomy and pathology had produced 'modern practitioners', noted the surgeon Samuel Cooper in his 1822 edition of *Dictionary of Practical Surgery*.[96] In London and Edinburgh, surgeons began to extend the range and complexity of procedures and from around 1820 public interest in operations seems to have increased.[97]

It was during this period that some surgeons began to think of surgical pain in a new way. As elsewhere in medicine, pain had long been understood to perform a critical role during operations by sustaining the body's vitality whilst its systems were being depressed by the action of the knife. But although opiates were efficacious in the treatment of chronic and terminal pain and were used to treat post-operative pain, James Moore, surgeon at St George's hospital in London at the end of the eighteenth century, concluded that 'the strongest dose we dare venture has little or no effect in mitigating the sufferings of the patient during the operation'. For this reason Moore began to consider other ways of effecting insensibility and designed a steel contraption which was intended to compress the principal nerves, thus diminishing the sensibility of the limb prior to amputation. It achieved little success as patients complained that the pain of compression was considerable.[98] Other novel and localised techniques such as the use of moxa (burning the skin) and acupuncture were recorded but with little success, and attention turned to physical states characterised by the suspension of sensibility.[99]

In 1819, James Wardrop, lecturer at the radical Aldersgate School of Medicine, was consulted by a young woman with a tumour on her head. She was unable to remain still enough for surgery despite being held down, so Wardrop experimented by creating a state of syncope – a low pulse and intermittent consciousness – through bleeding prior to the operation. It was, he later claimed, no different to the state to which patients were frequently brought in severe operations through loss of blood and mental agitation. Operating on a 'bloodless' body was analogous to dissection; it improved visibility and removed the psychological pressure exerted by a distressed patient. No matter how courageous the patient before the operation, explained Wardrop, 'many... often feel disappointed, as regards the degree of pain, and the time occupied in performing it, and the mind becoming fatigued and irritable, the sufferer has no longer the power of controlling himself and is unable to remain steadily in the same position'.[100] He recommended using opiates pre-operatively (specifically laudanum because it lessened the risk of vomiting) but observed that however 'free' their use, the 'mental inquietude' of the patient tended to negate their effect. Wardrop's first attempt at inducing syncope was made in 1819 and appears in a later series of lectures on surgery published in the *Lancet* in 1833. Despite his optimism there appears to have been little take-up of his suggestion as a strategic method for operative pain-relief though he noted that the creation of syncope was a 'common rule of surgery' before attempting to reduce dislocations or hernias – procedures indeed where there was no fear of blood loss.[101]

Thus from around 1820, surgeons like Moore and Wardrop began to seek ways of avoiding surgical pain. This is also the period when a new breed of experimental physiologists developed the new anatomical view of the body by experimenting on animals to analyse the functions of different parts of the nervous system. It is in this context that we can understand the animal experiments on asphyxia by the Shropshire surgeon, Henry Hill Hickman, and the work in London of the physician and physiologist, Marshall Hall. Both had links to Paris, which was the centre of the new approach to nervous functions.

## Hickman, Marshall Hall and the suspension of life

Trained in Edinburgh at a time when the medical school was at its peak, Hickman had sat the exams of the Royal College of Surgeons and joined the Royal Medical Society of Edinburgh, before establishing himself in practice in Shropshire in the early 1820s. He described himself as a surgeon and there is nothing to suggest that his experience was out of kilter with that of other provincial surgeons whose practice was more likely to consist of treating patients with leg ulcers and chronic infections than performing major operations.[102] The first description of his work on 'suspended animation' can be found in a letter of 1824, addressed to his local 'man of science', Thomas Andrew Knight, Fellow of the Royal Society.[103] Knight was well known for his work on plant physiology and in the early 1800s had collaborated with Davy on research into plants in vegetation, but there is no evidence to show that Hickman's work had been shaped by this connection.[104] Rather, Hickman wanted to ease the suffering of surgical pain. His letter described how he had created an artificial state of 'suspended animation' in animals through the inhalation of carbon dioxide gas, respired air and excluded air. When respiration had ceased, he amputated limbs and removed ears and tails before the animals regained sensibility. He suggested that inflating the lungs with bellows or galvanism could be used to restore the 'powers of life', although the animals he experimented on had all recovered naturally when exposed to fresh air. He also noted that inhalation of carbon dioxide gas appeared to limit haemorrhage and aid wound healing.[105] Later that year he printed a short pamphlet which included a second, slightly fuller letter addressed to Knight, indicating that Knight and others had asked him to present his work formally. It is possible that the intention was for Knight to present it to the Royal Society but its transactions carry no record of this. His letter was published in the Shrewsbury papers and in the *Gentleman's Magazine*, though the editor of the latter was dismissive: 'it may be doubted whether the pain of his operation, and especially in the recovery, would not equal or perhaps surpass that experienced in the usual mode of operation', he wrote.[106]

Four years later, Hickman journeyed to Paris and during the several months he spent in the city, he sent details of his experiments to Charles X, asking for help in pursuing his research in the French schools. His letter was forwarded to the Académie de Médicine and a committee set up, but apart from Larrey, chief surgeon during the Napoleonic wars, no other surgeon expressed interest in the proposal.[107]

This is a slim set of sources certainly, but telling enough to facilitate a sketch of Hickman's model of the body, and the way in which it connects to the earlier model of Davy and the later one of the 1840s. First, and most crucially, it accords no function or purpose to the presence of pain during surgery. Rather, Hickman suggests that the 'best effects' of the process 'would be produced by the patient's mind being relieved from the anticipation of suffering, and his body from the actual suffering of a severe operation'.[108] Whereas Davy's model was monistic, Hickman's is dualistic and supports a separation between the functions of the mind and those of the body. It reflected the new work of the 1800s on the nervous system and the changes this had brought about in understandings of sensibility and the process of death.

In 1800, Xavier Bichat's research into the states of life and death depicted death as the gradual sequential elimination of the functions of the organs of the body.[109] For example, when respiration was suspended chemically – perhaps through breathing non-respirable gases – the functions of the brain would first be interrupted, followed by the cessation of sensation, locomotion, and the working of the lungs, then the action of the heart and the circulation would be annihilated, succeeded by processes like secretion, exhalation and digestion, until finally the body would lose its animal heat and death would be final. His model of the body comprised two types of life: animal life, seated in the brain which supported the higher functions of sensation, perception and volition; and organic life which involved the vegetative nervous system, digestion, circulation and so on. Whereas concussion, haemorrhage or asphyxia would extinguish animal sensibility, organic sensibility would often survive such incidents.[110]

The 1810 atlas of Franz Joseph Gall confirmed the brain as the organ of the mind and construed all mental phenomena as functions of organised matter.[111] But it was the anatomical investigations into the brain during the 1810s – in Britain by Charles Bell, and in France by Francois Magendie – that localised functions and gave rise to the possibility that sensations could be disassociated from the vitality of the body. Bell and Magendie independently linked specific sites to specific functions and distinguished the different sensory and motor functions of the anterior and posterior spinal nerve roots.[112] Rather than all nerves having the power to convey sensation or excite movement, suggested Bell, they were 'as distinct in office, as they are in origin from the brain'.[113] The French physiologist Marie Jean-Pierre Flourens later confirmed the specificity of different nerve fibres in a series of experiments performed on pigeons, published in the same year as Hickman's experiments. Flourens showed how a pigeon losing both cerebral hemispheres became blind, losing one hemisphere caused blindness in the opposite eye and removing the cerebellum from a bird destroyed its balance although sight and hearing was unaffected.[114] He would later draw on this series of experiments to show the pathway of ether and chloroform through the nervous system, but in 1824 the most crucial aspect of his work for our purposes was the location of sensation and volition in the cerebrum, and not in the spinal cord. Hickman's proposition that the body would become insensible to the sensations of surgery if the mind of the patient could be 'suspended' becomes plausible therefore in this new context.

The second key difference between Hickman and Davy relates to their sense of control over the process of death which reflected the way in which understandings of asphyxia had changed since the 1790s. Hickman interpreted the phenomenon of 'suspended animation' as one example of asphyxia; a state in which the vital phenomena were suspended from some cause interrupting respiration, but in which life was not actually extinct and could therefore be restored. For him asphyxia was a key staging post along a process of death: he was confident that he had the power to halt the process and restore life if necessary. In one experiment lasting 17 minutes, he used bellows to occasionally animate the dog whose leg he amputated. He also suggests the use of galvanism as a restorative of life.[115]

By 1824 then, Hickman felt strongly enough about the negative effects of surgical pain to suggest that a temporary state of asphyxia could be artificially created to suspend the life of the mind and remove sensibility. The mesh with Beddoes and Davy can be found primarily in his overall approach which was descriptive and analogous, rather than analytical. He does not, for example, offer any explanation of the means by which carbon dioxide suspended respiration, nor does he attempt to contextualise his work in the wider fields of chemistry or physiology. Indeed, his distaste for vivisection suggests that he curtailed his practical work as far as possible. Unlike Marshall Hall, his analysis of the animal body was focused strictly on the development of a clinical technique. He failed because the remedy seemed worse than the problem. An anonymous letter to the *Lancet* in 1826 robustly condemned Hickman's proposal for the way in which it suggested that 'a man who was about to have a tooth drawn' should 'be previously hanged, drowned or smothered for a few minutes in order that he may feel no pain during the operation'. The sensations of such a process, continued the correspondent, would be 'far more horrible than the pain inflicted by ordinary operations' and even if a patient was made insensible of an amputation, the 'inflammatory' healing process would not be ameliorated.[116] Pain, it seemed, was impossible to avoid. Even if it could be stalled, sooner or later the body would have to tolerate the unpleasantness of returning sensations.

Hickman's dislike of animal experimentation was widely shared in Britain. From the 1750s, as part of the broader humanitarian movement, there had been protests against the use of animals on the grounds that frequent repetitions of experiments did not have the power to reveal any new knowledge and were purposelessly cruel.[117] Bell during his work on the functions of the brain showed equal repugnance to the use of animals and argued that many of his findings derived from induction rather than experiment.[118] A public furore broke out in 1824 when Magendie visited London and performed vivisection during public lectures.[119] Of course, such concern about animal pain is in itself further evidence of the new view of all physical pain as unsupportable.

But not all British doctors were dissuaded from animal experimentation. In the decade after Hickman's experiments, similar questions were elucidated by Marshall Hall, whose work joined with French studies to reveal a distinct sub-system of the nervous system which could support life without sensation or volition. Marshall Hall had trained in Edinburgh, Paris and Germany and taught at several of the radical private medical schools in London.[120] He began work on the nervous system and involuntary movements – reflex actions – around 1830, and having first noted the phenomena of reflex action in the tail of a newt, suggested that reflex action was seated in the spinal cord and the medulla oblongata and was independent of sensation. The nervous system might thus be understood at three levels: the brain which supported sensation and volition, the spinal cord which supported reflexes such as swallowing, and the vegetative nervous system which supported respiration, circulation and digestion. The separation of these systems and functions was a matter of much technical debate, but all such work was anathema to those who understood all the body's processes to be dependent on a vital spirit, which in man was related to the soul. Many conservative doctors despised physiological analysis as mechanistic and atheistic, but it was supported by radical practitioners.[121] Robert Grant, professor of comparative anatomy at London University, taught Marshall Hall's theory to medical students during the 1830s using insects to demonstrate reflex actions.[122] Richard Grainger published a study of the function and structure of the spinal cord in 1837, and by 1840, the *British and Foreign Medical Review* could draw attention to the way in which the knowledge of the nervous system had been revolutionised since the 1810s.[123]

By 1840, Marshall Hall's theory had provided the framework for understanding how the functions of respiration and circulation could exist in a body, independent to those of sensation and volition. It did not, however, convey how such a state could be safely and surely attained, and for surgeons this problem was becoming ever more pressing as surgical innovations were exacerbating the problem of pain.

## Conservative surgery and self-control

Whereas surgeons practising c.1800 had understood amputation to be the safest form of intervention, from the 1820s onwards there was a shift to preserve limbs and tissues through the excision of diseased or injured bone and tissue. Conservative surgery, as it became known, was a mark of the 'modern' surgeon who regarded amputation as a last resort.[124] Elite surgeons such as James Syme in Edinburgh and William Fergusson in London pioneered new procedures such as the excision of the foot or the elbow in an attempt to retain the remainder of the limb; tumours of the breast were removed using similar techniques.[125] But the consequence of this new approach was that it prolonged considerably the period of suffering. Whereas an experienced surgeon could amputate a limb within seconds, Stanley notes that the removal of breast tumours could last between 30 and 90 minutes.[126] Surgeons were quick to emphasise the utilitarian advantages of conservative surgery. When Robert Liston excised the carious bone from a young woman's foot, at University College hospital in 1844, he drew attention to the way in which the procedure had saved the patient her whole foot; it was 'an unequivocal example of the powers and advantages of true surgery'.[127] Liston was very conscious of the problem of pain and attempted to adapt his techniques, suggesting that: 'The . . . parts should be divided by a single incision, rather than that the patient should be tormented . . . by a slow and tedious procedure, bit by bit . . . incisions from within outwards . . . give much less pain than those in the opposite direction.'[128] But the accounts of those who underwent operations during this period suggest that such measures did little to mitigate the acute sufferings of surgery. The experience of George Wilson, who described the operation on his foot by Syme in 1843, has been frequently quoted as evidence of the enormous physical suffering of pre-anaesthetic surgery. But besides the appalling mental and physical sensations he endured during the process, he took the trouble to note that the novelty of the procedure had the benefit of leaving him with a 'more useful limb'.[129]

Surgery for remedial rather than life-saving reasons also began to be performed. In the 15 months between September 1840 and December 1841, new operations for stammering, squinting, club foot and cataract appeared in the *Lancet*, suggesting that contrary to much historical interpretation surgical development was not dependent upon the innovation of anaesthesia.[130] However, although surgeons had the anatomical knowledge and the technical skills to perform more complex and technically demanding procedures than 50 years earlier, and patients were receptive to the benefits of such innovation, operations remained the last resort of surgical practice. Most surgical patients were treated with therapies or local applications – poultices to reduce inflamed joints, or bandaging procedures to heal fractured or diseased limbs. This approach stemmed from the belief that doctors should facilitate and support the body's own powers of healing, rather than intervene and upset the balance of nature; it reflected the seemingly inescapable problems of pain and post-operative infection. Thus by the 1840s, there was a clear tension between surgical ability and patient endurance.

Alongside this emergence of conservative surgery, surgeons adopted a new strategy and sought to shape the patient's response to the pain of surgery through the influence of the mind. In 1818 during a lecture at the Royal College of Surgeons, Anthony Carlisle told his audience that the moment had come for 'the introduction of moral influence to mitigate or arrest the sufferings of surgical patients'.[131] Earlier generations of surgeons had encouraged patients to vocalise sensations as a way of expelling pain, just as purging or bleeding voided unwanted or excessive fluids from the body.[132] In 1811, Henry Cline advised a patient undergoing lithotomy that he should cry out rather than depress his energies by trying to control his reaction and in the same year, Fanny Burney had screamed throughout 'the whole time of the incision' as 'the dreadful steel was plunged into the breast' during her mastectomy.[133] But the shift in the construction of pain from one of physiological purpose to that of needless suffering promoted new tactics. Surgical textbooks depicted the pre-operative meeting between patient and surgeon as a key opportunity for the surgeon to mentally prepare the patient to 'meet the evil' of the operation. Confidence and honesty on the part of the surgeon about the actualities of the forthcoming operation – length of procedure, intensity of pain – were the most effective means of fortifying and tranquillising the patient's mind.[134] However, as Wardrop noted, in many cases, the 'strength' of the mind was not sufficient to control the patient's response to physical pain once the operation had commenced. And it was not uncommon, observed Thomas Curling, surgeon at the

London hospital, for operations to be abandoned 'owing to the impatience and want of self-control of the sufferer'.[135] The new stress upon self-control and fortitude can be discerned in operation reports of the 1830s and 1840s and approbation was given to those patients who were controlled enough to remain as silent as possible during surgery or at least display 'fortitude' by limiting their expressions to moans or groans.[136] In fact, wrote one correspondent to the *Lancet* in 1840, 'when a person chooses to die rather than submit to an operation, it is generally an evidence of deficient fortitude'.[137]

The stress on such qualities in the surgical context mirrored the way in which control of self and society featured high on the agenda of Victorian reform and was integral to all levels of social change. Indeed, I would argue that medical practice had directly reinforced such values through the wide use of opiates which were successful in easing the painful sensations of death, disease and childbirth and thus restoring to both patient and doctor a sense of control over the process. That such control was an aspiration of Victorian society is further emphasised by the broad social use of opiates to control the manifestation of emotions: Prime Minister William Gladstone was amongst many public figures who took laudanum to calm the anxiety induced by public speaking; babies were doped with Godfrey's cordial to prevent them crying; and unruly animals were calmed with opiates before going to market.[138]

It would seem then that the extensive use of opiates throughout British society had reconfigured expectations of the norms of behaviour in a wide range of social and medical contexts. But in the particular context of surgery, there was no effective means of alleviating surgical pain to the extent that by 1840, French surgeon Velpeau expressed the view of many doctors that finding a solution to the pain of surgery was a 'myth'.[139]

## Mesmerism

When considering the contexts of major discoveries, we are always at risk of 'presentism'; of doing 'history by hindsight'.[140] Our knowledge that the use of ether in 1846 marked the beginnings of modern anaesthesia makes it all too easy to construct its earlier history as a sequence of events which created a predestined space for ether; to imagine surgeons and others as looking for anaesthesia as the solution to the problem of surgical pain. But this cannot be the story for ether anaesthesia; we know that it was discovered by a marginal practitioner – a dentist, in a marginal country – the United States of America, and we shall consider the reasons below. And when one man succeeded, others revealed their earlier efforts, and by no means do they add up to a sustained assault on a pressing problem. Yet, as this chapter has argued, the issue of surgical anaesthesia, which in some senses was ages old, had been newly defined and highlighted by the 1840s. The surgical range had increased substantially, and conservative techniques had prolonged operations and the suffering involved. The increased use of opiates was indicative of a decreased tolerance of pain, and the injunctions about self-control suggest a decreased public tolerance of suffering freely expressed. Surgical pain had lost its rationale in medical theorising, and the body could be envisaged as capable of unconscious life, where vital functions would continue in the absence of feeling.

The new prominence of the problem is further evidenced by the debates in the early 1840s over the place of mesmerism in surgery. Few people knew about Hickman's work, and no one was thinking back to Davy, but the whole British medical world knew that the highly controversial techniques of mesmerism had been tried in operations. For the first time, elite surgeons publicly attempted to operate on unconscious patients; indeed it was the mesmerists who gave new meaning to the word 'anaesthesia'.

Anaesthesia was listed in Cullen's nosology of the 1750s and understood to be a cluster of diseases in which the key symptom was the loss of touch. By 1839, anaesthesia was understood as a condition arising from a diminution of the sensible power of the nerves. It could be caused by local conditions – the pressure of a bandage for example – or it could affect the whole body as in cases of palsy, apoplexy and hysteria. And, the return of feeling to the body was accompanied by unpleasant sensations.[141] The *Lancet* had published a series of lectures by the French surgeon, M. Andral in 1833, on diseases of sensation; anaesthesia with its symptoms of diminished sensibility was at one end of the scale, at the other with a heightened capacity to feel was hyperaesthesia.[142] By the 1840s, the new practices of mesmerism and hypnosis endowed anaesthesia with new meaning and it came to stand for a bodily state in which sensation and volition were suspended.

The history of mesmerism in British culture during the mid-nineteenth century has been powerfully told by Winter.[143] An earlier form of the practice – animal magnetism – was popular in the 1790s but had become politicised and marginalised throughout Europe in the wake of the French Revolution. Like Davy's work on gases, it was suppressed in the conservative political climate which accompanied the Napoleonic wars. A revival began in France soon after 1800 and by the late 1830s in

London, individuals such as Herbert Mayo, professor of anatomy and physiology at King's College, and John Elliotson, professor of practical medicine at University College were captivated by the powers of the mesmeric trance to create a bodily state in which volition and sensations appeared suspended. Elliotson was a radical reformer, receptive to all innovation and was particularly keen to explore the contention that a trance could effect the displacement of sensibility to a different part of the body. With the intention of establishing mesmerism as a useful medical tool, he undertook a series of experiments upon the O'Key sisters at University College in 1837 which ended in a debacle when Thomas Wakely, acerbic editor of the *Lancet*, proved that the effects derived from 'mere imagination' and that Elliotson had been duped. From this point, the *Lancet* mounted an intense campaign against mesmerism in which it was constructed as bearing the very worst attributes of quackery and sham. Reports of mesmeric anaesthesia being used successfully in surgery were also dismissed although there had in fact been considerable success with this technique. The first mesmeric operation to gain substantial publicity took place on a 42-year-old labourer, J. Wombell, in 1842. The mesmerist was a reputable barrister, William Topham, who spent several days ahead of the operation putting Wombell into repeated trances. Whilst W. Squire Ward, surgeon at Ollerton Infirmary in Nottinghamshire, amputated Wombell's leg, he remained motionless even as the sciatic nerve to the spine was cut. The dispute which broke out focused on the key area of contention: were the effects real or imagined? The evidence used by opponents such as Wakely, and the surgeon Robert Liston, drew on Marshall Hall's theory of reflex action. They argued that if Wombell had been insensible to the operation, his limb should have shown a reflex jerk at the point when the sciatic nerve was cut. Elliotson countered this point; mesmerism was a 'nervous phenomenon' which negated any usual action in the nerves, such as the reflexes, he suggested.[144] Despite such intellectual controversies, many surgeons, who perceived the pain of surgery to be Bentham's 'inherent evil', were open to experimenting with mesmeric anaesthesia. James Simpson, who was later to become the best-known advocate of chloroform, attempted to use it, as did several of the American dentists who later pioneered nitrous oxide and ether.[145] Despite such receptivity, mesmeric anaesthesia never meshed with British social and medical expectations of surgical pain-relief as ether was to do. The fact that it became a highly effective technique in the colonial context of India, used by the Scottish surgeon James Esdaile, is indicative of the contextual dependency of innovation.[146]

Mesmeric anaesthesia did however provide evidence that it was possible to effect a state in the body in which sensibility and volition could be disassociated from the central functions of circulation and breathing. Nor was it an isolated example of constructing a model of the body in this way, as shown by the work of the Manchester surgeon, James Braid, on hypnotism. After watching the famous French mesmerist, Lafontaine, at work in 1841, Braid began to experiment with its possibilities. He tailored the mesmeric trance to produce a state which he described as 'nervous sleep' and which is historically accepted as the beginnings of the practice of hypnosis.[147] He explained the trance as a consequence of the voluntary suspension of the will on the part of the patient, rather than the enforced suspension through the magnetic passes and powers of the mesmerist. His explanation was more analytical than that offered by the mesmerists and integral to his construction of the phenomenon was the active participation of the patient. Indeed, this may have been the reason why, unlike mesmerism, hypnosis became an established, orthodox technique from the 1870s onwards.[148] Braid understood 'nervous sleep' to be a means through which the nervous system could be thrown into a 'new condition' which could prove useful for the cure of certain disorders. It was a process, he explained, which at first intensified the senses (he compared it to the effects of opium and wine), and consciousness often remained. In the second state, the senses passed into 'the most profound torpor' and there was no sensibility in the body.[149] The state of 'nervous sleep' was used successfully by Braid to provide pain-relief for minor operations, as well as the treatment of chronic diseases such as rheumatism and palsy. It seems likely that Braid had absorbed the implications of Marshall Hall's work on the nervous system which proved that sensation and volition were seated in the mind rather than the spinal cord. His work also suggests a familiarity with the nuances of a debate which broke out between Marshall Hall and Johannes Muller on the manner in which, through the emotions, the mind was capable of overriding certain reflex functions such as respiration; Marshall Hall later accepted that 'the influence of emotion is, indeed, both diffuse and extreme'.[150] For although Braid found hypnosis to be a highly effective method of pain-relief for the extraction of teeth and other minor procedures, he believed that knowledge of the forthcoming event in the patient's mind 'may render it impossible for him to become hypnotised deeply enough to render him altogether insensible'.[151] Practitioners, he advised, should obtain consent from the patient for the operation 'at some unspecified time', rather than allowing them to know in advance that the procedure was going to happen.[152]

For hypnotism, as for opiates, the patient's fear of the operation was a major obstacle to pain control.

By 1846 then, not only had a social receptivity to the control of sensibility at all levels been established through the increasing use of opiates, but the new anatomical and physiological constructions of the body showed how it was perfectly possible for life to be sustained in a body devoid of sensibility. This understanding is vividly expressed in Elliotson's oration to the Harveian Society in June 1846. His aim was to mount a defence of mesmerism but his belief that a bodily state in which sensation and volition were suspended without compromising the functions of respiration and circulation was physiologically sustainable is clear:

> A body of facts is presented to us not only wonderful in physiology and pathology but of the very highest importance in the prevention of suffering under the hands of the surgeon and in the cure of disease. The chief phenomena are indisputable: authors of all periods record them, and we all ourselves witness them, some rarely, some every day. The point to be determined is whether they may be produced artificially and subjected to our control: and it can be determined by experience only. The loss of common feeling, – anaesthesia, is but a form of palsy, and in it wounds give no pain. If this condition can be induced temporarily by art, we of necessity enable persons to undergo surgical operations without suffering. Whether the artificial production of these phenomena, or the performance of processes which so often induce them, will mitigate or cure disease can likewise be determined by experience only.[153]

Yet, despite the receptivity of British doctors to the artificial suspension of sensibility, modern anaesthesia did not emerge from this milieu. For the twist in the tale we turn to America, to gas sniffing, money, teeth and pain.

## 'Teeth extracted without pain'

The fact that it was an American dentist – William Morton – who established the innovation of ether, rather than a surgeon, has attracted little historical attention and has been related as one of the odd quirks of the history of anaesthesia. The focus of enquiry has remained firmly upon the issue of priority of discovery, rather than its context.[154] Pernick draws attention to the disputes over the status of dentistry – was it trade or profession? – and suggests that Morton's attempts to patent his discovery inflamed the debate.[155] But there is more to be said.

Since the 1800s, the power of gases to change the physical and mental states of those who inhaled them had been established as a popular social entertainment in fairgrounds and travelling shows on both sides of the Atlantic. For gases, as for mesmerism, the amusements of the radical elites of the 1790s had become common playthings. Audiences gathered to watch individuals, intoxicated with 'laughing gas', behave in a disinhibited manner with scant regard for the social norms of behaviour. That such a spectacle became construed as social entertainment is highly suggestive of the stress that was placed upon self-control in society; the humour lay in the apparent disregard by such individuals for such norms. So too in medical schools, students were often given the opportunity to breathe nitrous oxide in chemistry lectures and again, the behaviour of exhilarated students was a sight enjoyed by their peers.

It was indeed in this popular culture of amusement that the first surgical operations were performed under ether, without attracting much attention. In the village of Jefferson, Georgia, in the American South, a local doctor was asked to make some nitrous oxide for a group of young men. Not having the apparatus, he gave them ether instead, and this became a local fashion. One of the local ether sniffers then needed a minor operation for the removal of a cyst, but he was unusually fearful. The doctor, Crawford Williamson Long, knew from his own experience that ether often removed the sense of pain, so he suggested the operation be performed after ether breathing. It was successful, and Long continued the practice on the one or two patients a year in which he believed the inhalation of ether to be applicable, one of them a negro child.[156] He did not report his results until others had gained credit for a discovery which by then had swept the world. Why was his work not news?

Ether was first synthesised in 1540 and it became well known as an anti-spasmodic for asthma and as a useful solvent that evaporated quickly, one that could cool and numb the skin. It was known to produce excitation when inhaled, similar to that of nitrous oxide, but could also produce lethargy.[157] For Long it could be substituted for nitrous oxide as an amusement for bored young men, and it joined opiates and alcohol as a possible means of avoiding pain, including the pain of minor or rapid surgical operations. It could be used in patients who were uncommonly fearful, perhaps especially those who were already accustomed to ether breathing. Long was not concerned with physiology or chemistry and sought no financial gain or professional advancement; he was a country doctor far from major medical schools and the professional spectacle of major operations. Most of his patients were stoical Christians who believed in God's providence and who were no more likely to sniff ether for surgery than they were for amusement. Indeed, he was advised by some of his colleagues to abandon his experiments.[158]

For well-trained doctors, breathing gases was dangerous because beyond the initial stimulating phase they induced a process of death through asphyxia.[159] Medical students were taught that ether and nitrous oxide had narcotic and poisonous properties with the power to destroy sensibility and irritability. They may have tried a few whiffs of gas, they also witnessed the death of animals from asphyxia through their exposure to ether and nitrous oxide. Within this context it is understandable that there would have been little enthusiasm for exploring the further possibilities of gas inhalation. But the Boston dentist Morton was outside this culture. He was enough of a dentist to see considerable financial advantage in the new technique, and close enough to elite medicine to imagine the benefits of medical support for ether.

In 1846, Morton was only 26 years of age but trailed a history of unscrupulous business deals, debts and failed partnerships.[160] With little formal training he had set up in practice as a mechanical dentist specialising in the manufacture and fitting of artificial teeth. He had acquired his dental skills through an apprenticeship and although he had registered for 2 years of medical lectures at Harvard Medical School between 1844 and 1846, this seems to have been largely at the instigation of his future father-in-law, Edward Whitman, who was keen to ensure that his daughter, Elizabeth, should marry a 'regular' practitioner. During this period Morton continued to run his dental practice, so the extent of his attendance at these medical courses is unclear.[161]

By the early 1840s, improvements in dental technology meant that denture bases could be made out of a gold-alloy and this improved the fit of the dentures inside the mouth.[162] But many patients could not bear the pain of having rotten teeth and stumps removed before artificial teeth were fitted. Patients often, 'especially…delicate females', abandoned proceedings halfway through.[163] By the autumn of 1845, Morton's practice had become almost exclusively that of mechanical dentistry.[164] He was not alone in recognising that a method of ameliorating the pain of extractions would give his business a market edge: Horace Wells' motivations for experimenting with nitrous oxide in 1844 were exactly the same, although his attempt to establish nitrous oxide had resulted in a humiliating public failure.[165]

In many ways, dentistry was more supportive of experimentation than surgery. Extractions were not life-threatening procedures, nor did they subject the patient to much loss of blood. Dental culture was also far more commercial than medicine; it was common for new techniques and discoveries to be protected by patent rights or kept secret and it seems possible that the culture of American dentistry was more akin to commercial practices than British dentistry.

Morton's experiments with ether cannot be explained through a particular skill or interest in chemistry. His choice of ether derived from the suggestion of the chemist and geologist, Charles T Jackson, who also practised dentistry and employed ether as 'toothache drops'.[166] (Jackson later disputed Morton's claim to the discovery.[167]) In 1844, after learning from Jackson of the power of ether to numb the nerve of a tooth, he applied liquid ether to diminish the pain of a filling. This was not a particularly novel use of ether – *Hooper's Medical Dictionary* of 1820 cites the external application of ether as a cure for toothache and headache. Following this success, Morton borrowed Jackson's chemistry and medical books to ascertain that 'there was nothing new or particularly dangerous in the inhaling of ether'. It was, he said, a 'toy of professors and students', well known for its intoxicating and stupefying properties.[168] During the summer of 1846 he carried out some more experiments and determined that sulphuric ether was the most effective form of the chemical. He inhaled ether on a handkerchief and used it on household pets, including a water spaniel whom he held over a tin pan containing ether soaked into cotton wool until the dog 'wilted completely away in his hands'. After two or three minutes the dog became as 'lively and conscious as ever'.[169] On 30 September 1846 Morton was visited by a patient who was hoping to have his tooth extracted under mesmerism; Morton offered him ether and the tooth was extracted without pain. It was therefore a similar experiential model of experimentation, analogous to that of Davy's nitrous oxide work although at a far less sophisticated level. In no sense can Morton's experiments be constructed as part of the new analytical scientific medicine of 1840s Europe.

Morton's discovery of the power of ether to remove sensibility can be explained in part through serendipity; his first trials upon himself and others induced insensibility rather than asphyxia. But it was the strength of his ambition to exploit the commercial advantages of ether which sealed his success. His aim was to benefit from ether through patenting the new technique, not just for dentistry, but in surgical operations. He was canny enough to appreciate that this would only be possible if ether gained the full approbation of the Boston medical elite; medicine was perceived to be a higher authority for dentistry and for this reason dental innovations were frequently underwritten and certified by medical practitioners.[170]

The setting Morton chose to demonstrate the powers of ether was the Massachusetts General Hospital in Boston, a bastion of medical respectability which he would have known as a medical student. The surgeon, John C. Warren, was one of Boston's medical elite with an impeccable professional history. He had trained in Edinburgh and then held posts as Dean of the Medical School and as Professor of Anatomy and Surgery before becoming one of Boston's leading surgeons. He was a founder of the *Boston Medical and Surgical Journal* (later to become the *New England Journal of Medicine*), and had helped establish the Boston Medical Library. Warren had been party to Wells' failed attempt and his accedence to Morton's request suggests a keen receptiveness to the prospect of pain-control. The demonstration was staged before a group of elite individuals whose opinions carried enough weight to sway public and professional opinion. In the audience were Professor Jacob Bigelow, professor of materia medica at Harvard Medical School; his son, Henry Bigelow, surgeon at Massachusetts General Hospital; and Edward Everett, President of Harvard University and former United States Ambassador to London.[171] Morton was fully aware that it would be the approbation of such individuals that would secure ether's acceptance. That it was Morton's marginality to medical culture that freed him to take risks that doctors perceived to be impossible was later expressed by the New York surgeon, Valentine Mott, who commented that not even the most 'bold and adventurous' surgeon would have had the 'temerity' to experiment with ether in the way Morton had.[172]

Ether's acceptance in Boston was largely due to Morton's placing of the technique under the authority of the medical elite and once doctors had established its efficacy, news spread worldwide. Morton proved the bridge between self-experimentation, dental arts, commercial aspirations and the international medical elite. The next chapter will show how, within the context of British medicine, the artificial creation of a bodily state without feeling through the inhalation of ether was sustained and developed into modern anaesthesia.

# 2
# Altered States

From the moment that ether was first used by the London surgeon, Robert Liston, for the amputation of a thigh 'the future of anaesthesia was assured', notes one historical account.[1] It is an assumption shared by most other writers on the subject. Yet the reality of ether proves far more complicated. Chapter 1 established how by the 1840s, the high value placed by Victorian society upon self-control had created a social intolerance towards the expression of physical suffering – be it during death, surgery or childbirth. Just prior to ether, surgeons expected patients in the operating theatre to exercise restraint and fortitude when undergoing painful surgery. But ether made patients giggle, tell jokes, curse and struggle; it turned bodies dark red or purple, or reduced them to breathing corpses; and its pungent smell lingered on the breath and clothes. The sombre mood of patient stoicism against adversity was overriden by laughter and intoxication. This gamut of altered states threatened the propriety of surgical practice. If ether placed patients beyond individual self-control by intoxicating, poisoning or asphyxiating them, then how could reputable doctors accommodate it? Like John Snow, some doctors succeeded in establishing efficacious methods of producing insensibility but many found the gas unworkable and abandoned it after initial trials. Had it not been for the introduction of chloroform in 1847, anaesthesia may well have remained on the margins of practice for some time.

Chloroform established insensibility quickly and easily and resolved many of the dilemmas created by ether. But within weeks of its introduction, a different, but equally worrisome issue emerged: its propensity to kill without warning. In London and Edinburgh, doctors hotly disputed the mode of chloroform death – was it caused by asphyxia or poisoning? An elite minority placed the inhalation of ether and chloroform within the frameworks of the new scientific medicine. It became a universal process which produced a predictable sequence of responses in all bodies and was of potential benefit to most patients. The majority placed the new technique within the familiar therapeutic paradigms of biographical medicine. In their view, the individual constitution determined the specific effects of the agents; breathing gas was an unpredictable and risky process and the benefits had to be judged for each patient.

In this chapter we reconstruct the introduction of ether and chloroform anaesthesia to British medicine and Snow's swift establishment of a framework of principles for the new technique. The 'Yankee dodge' provoked a welter of debate and controversy that reveals clearly the attempts of doctors in the 1840s to reconcile the new scientific views of the body with the traditional principles of biographical medicine.

## This Yankee dodge!

Morton's shrewdly staged presentation of ether to the elite of Boston not only ensured its take-up throughout America, but also tapped into the international networks of the scientific and medical elite. Those who witnessed the first operations at Massachusetts General Hospital were convinced of the potential of the new technique and helped Morton by becoming strong advocates.[2] Henry Bigelow reported the event to a meeting of the American Academy of Arts and Sciences and gave a paper to the Boston Society of Medical Improvement, which was later published in both the *Boston Surgical and Medical Journal* and the *Boston Daily Advertiser*.[3] His father, Jacob, having witnessed 'limbs and breasts . . . amputated, arteries tied, tumours extirpated and many hundreds of teeth extracted, without any consciousness of the least pain on the part of the patient', sent details of the 'new anodyne process' to fellow American botanist and physician, Francis Boott, who lived in London.[4] Boott received the news on 17 December 1846 and transmitted the message to London's medical and dental communities so swiftly that the first tooth under ether was extracted on 19 December, and on 21 December, Robert Liston, the metropolis' premier surgeon, used ether during a thigh amputation at University College Hospital.[5]

The immediate trialling of ether reinforces the argument that by the mid-1840s, pain had become a pressing problem in both surgery and dentistry. Winter has explored the role played by ether in outlawing mesmeric anaesthesia; it was a new solution to an existing problem: 'Gentlemen! this Yankee dodge beats mesmerism hollow!', Liston is reputed to have declared at the conclusion of his first operation under ether.[6] Edinburgh-trained Liston was known as a bold and fearless operator and frequently courted controversy by his willingness to attempt new and radical procedures, such as a scapula excision in 1822.[7] He was also part of London's medical reform movement, participating in the establishment of the British Medical Association in 1836 alongside other radicals such as Robert Grant, professor of comparative anatomy at London University; Richard Grainger whose work on the spinal cord supported Marshall Hall's new constructions of the nervous system; and William Farr of the General Register Office.[8] Liston's responsiveness to ether can be understood in part because he possessed the necessary intellectual frameworks to rationalise the creation of insensibility. He had been deeply engaged with the controversies surrounding mesmerism and its promotion by his colleague and rival at University College, John Elliotson, and had dismissed mesmeric anaesthesia as a quack practice. It was a sham, he said, the patient simply feigned insensibility to pain. Winter suggests that he believed mesmerism had introduced unruliness, rather than control, into the operating theatre. It thwarted rather than sustained his surgical ambition of operating on 'disciplined' and 'quiescent' bodies.[9] Ether too could create mayhem but Liston's first experiences of operating on an etherised body were successful enough for him to believe in the potential of the new technique to create his ideal operative state. But not all London surgeons shared his perspective.

Benjamin Brodie, past President of the Royal College of Surgeons and one of the surgical elders of London practice, had also received the news of ether directly from America – from Everett whom he had met during his period as US ambassador in London. Brodie remembered the surgical lectures he had given during the 1820s when he had demonstrated ether's narcotic and poisonous qualities to medical students by giving it to guinea pigs. His unpromising recollection was that ether vapour sent the animals to sleep and then killed them. On these grounds, he immediately questioned the safety of the technique.[10] Brodie's response echoed the argument put forward by the American surgeon Valentine Mott that Morton's temerity in using a chemical which in surgical circles was a known narcotic and classified within pharmacopoeias as a poison could not have been perceived as judicious practice by any surgeon.[11] In London, Brodie's view certainly held currency. Ether, said the *London Medical Gazette*, was 'a strong narcotic... its vapour speedily produces complete lethargy and coma', a state which the journal read as a 'temporary poisoning in which the nervous system is most powerfully affected'.[12] Despite such warnings, no deaths occurred during the initial trials of ether. For both doctors and patients, the allure of painless surgery was strong enough to quell initial concerns and a wave of experimentation spread across Britain.

## Spreading the news

Ether's proven ability to remove physical pain made it a discovery of the broadest humanitarian interest. In all quarters of the press, ether was headline news. The *Lancet* extolled 'the remarkable perfection' of the new discovery and published an average of four articles a week on the subject during the first 6 months.[13] The *Medical Times* wrote of its office being 'literally inundated with details of new operations' performed under the effects of ether.[14] Details of the removal of a tumour from a young woman at the Westminster Hospital were carried by the *Morning Herald*, listing as spectators Lords Walsingham and Morton, Viscount Falkland, Sir Henry Mildmay and 'many distinguished foreigners'.[15] Reports appeared in the *Times* as well as in local publications such as *The West Briton and Cornwall Advertiser* and *The Plymouth, Devonport and Stonehouse Herald*.[16] And the news carried through medical networks: Liston sent inhalers to some of his former students who worked in the provinces so they too could experiment with the new technique.[17] By June 1847, ether was being trialled worldwide (Figure 2.1).[18]

Yet few of the early ether administrations had the 'remarkable perfection' claimed by the *Lancet's* editorial. Those who flocked to the London operating theatres expecting to see quiet and motionless bodies were as likely to have witnessed patients struggling, held down and moaning during the operation. Early in January 1847, spectators at St George's saw the first operation, on a 'weakly' lad of about 19 years old, abandoned because a combination of fright and coughing stopped the ether from working. Then they watched a robust young man inhaling ether before the removal of a finger. He appeared to 'suffer a good deal from it', turning 'rather purple' in the face, and several of the spectators declared that 'the ether was as bad as the operation or worse'. When William Cutler took the knife to his finger, he shouted and snatched his hand back 'so vigorously' that there was no doubt in the minds of the audience that 'he suffered pain as acutely as if no steps had been taken to deaden it'. The final operation, an amputation below knee, was on a young man who 'followed the advice implicitly' and the resulting insensibility was 'very satisfactory'.[19] Ether's propensity to excite and disinhibit patients could create ludicrous situations. The Irishman who had his leg amputated at the London hospital spent his time giving 'sly winks and facetious nods to those surrounding him.…forcing from the bystanders involuntary laughter, and converting

**WONDERFUL EFFECTS OF ETHER IN A CASE OF SCOLDING WIFE.**

*Patient.*—" THIS IS REALLY QUITE DELIGHTFUL—A MOST BEAUTIFUL DREAM."

*Figure 2.1* *Punch* celebrates ether's power to diminish the pains of domesticity. *Punch* (1847). Reproduced by courtesy of the Director and University Librarian, The John Rylands University Library, The University of Manchester.

that which to the poor fellow was a most tragical event into little short of a farce'.[20] Yet when ether did prove effective and create complete insensibility, the pallor and immobility of patients appeared corpse-like and alarming to onlookers. Ether's 'effect on the system was appalling', remarked the eminent surgeon and geologist, Gideon Mantell, who travelled from Sussex to see operations at St Bartholomew's.[21]

The uncertainty created by ether's paradoxical effects was mirrored in its Janus-like identity. On one hand, it was spoken of as a scientific and progressive technique, capable of endowing medical practice with the highest values of humanitarianism and, on the other, its powers were used to entertain and amuse social gatherings in the tradition of 'laughing gas'. Robinson, who had been the first in London to trial ether in dentistry, described how on 24 January 1847 he visited Francis Boott and a group of acquaintances, including Prince Napoleon Bonaparte III. One or two members of the party took the opportunity to have teeth extracted under ether, but most wanted simply to experience its effects. The Prince came round from the vapour saying 'he felt the strength of ten men', and Boott himself 'inhaled twice during the evening, and pronounced the sensations to have been "glorious"'. Robinson noted that those who had breathed ether often 'evinced an almost uncontrollable desire for more'.[22] It was far more reminiscent of the Bristol breathings of nitrous oxide than was suggested by its portrayal in the medical journals.

## The science of ether

For all the difficulties, there was no doubt that ether had the power to render patients insensible to surgical pain. It is not remarkable then that ether was given unprecedented coverage by the medical press. What is notable is the broad acceptance that breathing ether was of scientific origin and therefore of an entirely different nature to mesmerism. From the first, the Yankee dodge had been placed firmly within the medical domain. This was one reason Bigelow gave in support of Morton's attempts to patent the discovery; the protection of the public from its use by unqualified practitioners.[23] The *Boston Medical and Surgical Journal* remarked early on that ether was 'based on scientific principles and is solely in the hands of gentlemen of high professional attainment', unlike the 'farce and trickery of mesmerism'.[24] These claims were replicated in British journals where editorials stressed ether's simplicity and accessibility; all doctors, by virtue of their professional status, had the power to dispense ether's humanitarian benefits.

As Chapter 1 noted, although many surgeons had been receptive to trialling mesmeric anaesthesia, it failed to become established, despite the pressing problem of surgical pain. Winter has argued that ether anaesthesia succeeded in displacing mesmeric anaesthesia because the technique complemented the 'social relations' doctors aspired to establish with patients and evidenced a clear medical authority over the body. Ether fulfilled the criteria of orthodox as opposed to unorthodox practice.[25] Indeed, this aspect was an important reason for the acceptance of ether by doctors: Liston's appreciation of the quiescent nature of an etherised body reveals such dynamics. But to understand more fully the differing responses to ether and mesmerism, on the part of both doctors and patients, we need to build on Winter's work by incorporating the new medical constructions of the body and public understandings of science.

We can begin by asking on what criteria was ether accorded a scientific pedigree? Winter argues that ether's pedigree was remarkably similar to that of mesmerism and in popular culture the use of 'laughing gas' and 'ether frolics' was indeed a well-established form of public entertainment.[26] She also suggests that 'etherists could not...distinguish it [ether] from the practice on which it was based: mesmerism'.[27] But this argument cannot be applied to elite doctors whose anatomical and physiological constructions of the body made feasible 'life without feeling'. For them, the state created by ether was intellectually defensible, even though its modus operandi could not be immediately explained. One of the difficulties with mesmerism was that its phenomena appeared impossible to 'place' within these new frameworks of knowledge. Hence the concern of those such as Liston and Marshall Hall that Wombell did not demonstrate any reflex movement during the amputation of his limb. In the context of the new scientific medicine of the 1840s, if physical states could not be linked to physiological or chemical origins then this was seen as proof that patient claims were either feigned, or a product of their imagination. That it was believed possible for a patient to suppress the physical experience of an amputation is itself highly suggestive of the significance invested in the powers of the self to control the body. This view is supported by the way in which the phenomenon of ethereal insensibility prompted some opponents to re-evaluate the status of the mesmeric trance. John Forbes, editor of the *British and Foreign Medical Review*, who had previously waged an intense war on the quackery of mesmerism, noted:

> if the new process shall supersede that employed, with a like object, by the mesmerists, we must concede to them that it supplies, from analogy, additional reasons for believing in their statements in regard to the production by their process, of insensibility to pain.[28]

For doctors who practised outside the new frameworks, explanations were far more likely to focus on the agent rather than the state it created. Like Brodie, many doctors drew analogies between ether and states of intoxication, poisoning or asphyxia. Nevertheless, at all levels the collateral sciences of chemistry and physiology were the bodies of knowledge within which ether was ratified in a manner that had proved impossible to sustain with mesmerism. But it was not just doctors' expectations of 'scientific' medicine that ether fulfilled, but also those of patients. Ether caught the public imagination in a way that mesmeric anaesthesia had not, and to explain this we need to turn to mid-nineteenth-century constructions of science.

By the late 1840s, 'science' had become established in the public domain as a norm of truth and a catalyst for progress.[29] Throughout all levels of society, the artefacts of science – steam, industrialisation and electricity – had transformed daily life. Scientific pursuits, in a myriad of forms, had become synonymous with aspirations of individual and social progress and were followed across the social classes – from the workers in the industrial cities of the north who used their sparse leisure time to scramble on the moors and collect natural artefacts, to the upwardly mobile middle classes of the metropolis who flocked to the Royal Institution to watch Michael Faraday demonstrate his work on electricity. For Joseph Gutteridge, a weaver in Coventry in the 1840s, his 'greatest triumph' was the construction of a microscope which he used to 'understand the various phases of matter and the myriad forms and functions of the animalcular existence'.[30] Gutteridge understood 'science' as a 'norm of truth' in a potentially harsh and unstable world.[31] Workers in the factories of Manchester were told that science would make them 'more skilful, expert and useful'.[32] For Victorians, science was not abstract or elitist, rather it held meanings of individual self-improvement and rational benefits; it was one of the foundations of their 'age of progress'. The characterisation of ether inhalation as a 'scientific' technique gave it an undisputable legitimacy and from a patient's perspective, the chemical, the flask and the breathing tube were all visible artefacts of its chemical nature.

In contrast, mesmerism was effected through the creation of an intimate relationship between patient and mesmerist.[33] Sitting opposite the patient, the mesmerist used hands and arms to make sweeping 'passes' over the patient's body, so close that the heat of the hands could be felt. In this way the mesmerist was understood to transfer his mental powers to the patient and produce an 'altered' state of mind. There were also suggestions that the encounter involved the transfer of 'vital fluids'. The process was explained in terms of the patient's will being subjected to that of the mesmerist. In this way physical sensation could be suspended and mesmeric subjects could undergo surgery with no apparent pain. But the contentiousness of mesmerism derived from the very way in which the process appeared to undermine the values of self-control; its success depended on the subjugation of will. It drew on supernatural practices and was understood to 'cast a spell over the man, and over his spirit'.[34] For the novelist Elizabeth Gaskell, a Unitarian who understood the findings of science as rational evidence of God's handiwork, mesmerism appeared a dubious practice with a propensity to twist minds.[35] Mesmeric anaesthesia was associated, said the *London Medical Gazette*, with 'selected cases, darkened rooms ... mystery in its employment'.[36]

Whereas mesmerism manipulated the self through control of the mind, ether mediated its effects through a chemical process; it could be read as a 'useful' extension of the popular frolics, a rational process rather than psychic domination. For many commentators, ether became one of the most tangible representations of the wider humanitarian social reforms – like slavery and prison reform, for example. For others it stood as one of the new powers of the age, comparable to the discovery of steam which had formed the basis of modern industrialisation processes. An editorial in *Littell's Living Age* equated the practice of a surgeon employing ether with that of a captain of a steamship. In the operating theatre, ether gave the surgeon the power to 'command the sensibilities of his patient'; at sea, the captain manipulated the power of steam – increasing it, slackening it and so on – in order to navigate his ship through a difficult channel.[37] Such examples suggest that the Victorian public saw the powers of chemistry in a very different way to the powers of another human mind.

By the end of January 1847, all the major London hospitals had trialled ether, and so had many provincial centres; there are records of at least 80 surgical operations using ether during these weeks.[38] (The technique had also been extended to animals; to relieve pain during operations on 'the brute creation'.[39]) It is a strong contrast to the take-up of mesmeric anaesthesia, which entered British practice in 1841, yet by 1846 seems to have been used only in 60 or so surgical operations.[40] Ether inhalation was not pleasant – patients complained of its pungent smell, feelings of suffocation and irritation to the throat; nor was it always successful – some patients refused to relinquish control over their sensibilities by breathing the gas.[41] A young woman, who had lacerated her perineum during childbirth and had already undergone surgery, agreed to a further operation 'if it might be done without causing her pain'. She disliked the process of inhaling so much: 'she would rather submit to the pain she had formerly experienced', she declared.[42] But the vast majority gladly traded its discomforts for the benefits of painless surgery. The first patients to be given ether in King's College Hospital were chosen by the surgeon, William Fergusson, precisely because their fear of pain had caused them to previously refuse operations. A middle-aged man who was 'very averse, for fear of pain', consented to circumcision to relieve the phymosis he had suffered from for the last 11 years. The ether worked well: 'he said he was not conscious of anything having occurred'. Ether also fulfilled the surgeon's promise to a 21-year-old man suffering from a crop of warts on the glans and lining of the prepuce, that 'they may speedily be got rid of without pain'.[43] Patients asked for treatment that they had previously been unable to contemplate. Mrs B visited the dentist Robinson, who had carried out the first tooth extraction under ether, 'desirous' of having 'thirteen different teeth and stumps removed by aid of the ethereal inhalation, if it could be effected *without pain*, but would not submit to the operation unless this could be guaranteed... nothing could induce her to submit to an operation performed in the usual manner'.[44] But although patients like Mrs B accepted the new technique without questioning its mechanism, doctors found that explaining ether's effects within medical frameworks was problematic.

## Meanings of ether

Because ether inhalation was classed as a new therapy its use could be shaped by the principles of biographical medicine.[45] These construed the body as a holistic system and used combinations of stimulants and depressants to balance its equilibrium. The first patients to experience insensibility were often given brandy, either in conjunction with the ether or afterwards; its stimulating effects were believed to counteract the depressive ones of ether.[46] The advice given on the administration of ether replicated that for other therapies such as opiates or poisons, where a 'dose' could have a range of effects on different patients. The creation of excitability, rather than insensibility, was explained by the individuality of different constitutions which sometimes caused a depressant (like ether) to act as a stimulant. The dentist Robinson urged those using ether to

> note every symptom or peculiarity, however minute, that may occur in the cases submitted to its operation... [and] to weigh well the previous habits and present state of their patients before they exhibit it; keeping in mind that there are peculiarities of habit and constitution, and particular states of health, in which the effects may be questionable, and some in which it would not be advisable to administer it.[47]

Failure to establish insensibility, or cases where the patient remained semi-conscious for some considerable time afterwards were also explained in terms of constitutional factors. This strategy stemmed from established practice but served to create a context in which the blame for effects other than insensibility was placed firmly upon the patient, rather than upon a lack of medical skill in administering the gas. But quantifying the effectiveness of the suspension of sensibility was far more problematic, primarily because ether created a state in which clinical impressions were at odds with patient testimony. This was a departure from established therapies in which efficacy was usually easily judged by particular and visible effects: an emetic caused vomiting, leeching produced swollen creatures which dropped off the patient's skin when replete, and cupping left indentations on the body. The trials of ether frequently presented onlookers with the appearance of a reality which was subsequently repudiated by the patient. In Boston, Bigelow, who had watched the first operations under ether, carried out some experiments on patients and noted how difficult it was to draw any conclusions independent of the patient's testimony because 'the phenomena of the lethargic state' were unreliable. Patients who displayed physical signs of distress such as frowning, moaning or struggling recovered without 'any knowledge of what had been done during their sleep' and no remembrance of pain.[48]

In London, a similar pattern was noted, and it jarred with the expectations of surgeons who were accustomed to using their privileged knowledge – gleaned from observation and examination – to diagnose disease and determine its treatment.[49] A timid, delicate-looking 26-year-old woman inhaled ether at King's College hospital whilst a large abscess on her buttock was drained: she 'moaned a little, and seemed as if in pain', yet when she was questioned afterwards, 'declared that she had felt none'.[50] Other reports spoke of the 'apparent insensibility' of patients who 'seemed to feel no pain'.[51] This lack of transparency in the workings of ether was problematic because it opened up the possibility that the effects of ether were creations of the patient's 'imagination'. Imagination in the medical context carried connotations of deceit and trickery as noted earlier in relation to mesmerism.[52]

Indeed, it would seem that it was because of a distrust of patients' accounts that many doctors inhaled the gas in order to quantify its effects for themselves. In Nottingham during January 1847, doctors gathered to watch Dr Gill, a physician of the town, inhale the gas. They arranged that Gill would stamp on the ground with his foot during the time he remained conscious. After about 4 minutes, the stamping ceased and Gill entered a state of 'perfect unconsciousness and insensibility'; his colleagues began to test its effectiveness by passing a needle through the skin on his hands and pinching the points of his fingers. The surgeon, Henry Taylor, who reported the details to the *Medical Times*, was explicit that the event possessed greater value for having been carried out on 'one of our own profession who is, therefore, better able to judge of the effects produced upon the sensitive parts of the system'.[53] Perhaps what Taylor could have better said was that it would have been more difficult for a doctor to have performed equivalent tests on a patient.[54]

However, although the innovation created new dilemmas it also resolved old difficulties. By the 1840s, many doctors and patients were disenchanted with the failure of the new scientific medicine to effect any significant breakthroughs in the treatment of sickness and disease.[55] The new anatomical and physiological work had produced new explanations of disease. But 'a knowledge of disease is not half its cure', a London medical journal warned its readers in 1830 and noted that prior to the 'cultivation of morbid anatomy in this country…the most eminent practitioners directed their whole attention to the discovery of remedial agents'.[56] From the perspective of the mid-nineteenth-century doctor, William Withering's discovery of the therapeutic benefits of the foxglove in the 1780s and Edward Jenner's work on smallpox in the 1790s appeared as shining examples of medical humanitarianism which were yet to be replicated by the 1840s generation. In this context, it is easy to appreciate why, once ether had been proven, it was immediately taken up as an innovation of enormous symbolic, as well as practical, importance.

It was doctors, rather than patients, who first became disillusioned with the new technique. Chitty Clendon, lecturer on dental surgery at the Westminster Hospital, wrote of his varying success in using ether for tooth extractions. In four out of nine cases, ether failed completely to relieve the pain, although the remaining cases were slightly more successful.[57] Even Liston, who had delighted in ether, suffered so many failures that he came close to abandoning its use.[58] The difficulties in many cases arose because of the technology of the inhalers (Figure 2.2). Despite the claims of scientific pedigree, the first inhalers were contrived in ways which took small account of chemical facts. Most replicated Morton's glass flask into which the ether was poured on a sponge. Questions as to the optimum conditions for the vaporisation of ether, for example, appear not to have been asked, even though the chemist, Joseph Black, had described the principles of latent heat and vaporisation in the 1790s.[59] Thus, many inhalations resulted in the

*Figure 2.2* Hooper Ether Inhaler. One of the earliest inhalers made in London during December 1846. Early Technology.

patient breathing in 'air much colder than the freezing point of water, and containing very little of the vapour of ether'.[60] On the operating table this produced an excited, uncontrollable patient rather than an insensible body. Ether was at risk of being 'reduced to the level of one of those pretended discoveries – to one of those scientific *puffs* which come periodically to amuse the curiosity of the public and satisfy its irrational passion for all things erroneous and untruthful', commented French physiologist Francois Magendie.[61] His concern was shared by the French surgeon, Boullay, who reminded members at a meeting of the Académie de Médicine in Paris on 26 January 1847, that although: 'The public is quite preoccupied with it [ether], and rightly so ... no one has yet determined even the proper dose of ether to be given'.[62] Neither Boullay nor Magendie were aware that in London, one doctor was attempting to ascertain precisely that.

## Snow and the principles of anaesthesia

In 1846, Snow was working as a general practitioner in the Soho area of London (Figure 2.3).[63] He had come from a working-class background in York and entered medicine through the apprenticeship system, training in Newcastle upon Tyne and working as a general practitioner's assistant before moving to London to study for the exams of the Society of Apothecaries and College of Surgeons. Once qualified, he bucked the trend by setting up in practice at 54 Frith Street, on the west side of Soho Square; the majority of doctors without family or social connections to the capital to support them returned to their home or apprenticeship towns as it was easier to make a living.[64] Professional success in the mid-nineteenth-century London medical market was dependent upon patronage or other entrees into the elite networks.[65] Snow enjoyed neither. Instead he was to gain his reputation within London medicine by wedding science to the practice of anaesthesia. Such an approach could only be sustained in the context of the new view of the body which focused on the universality of systems, tissues and organs, rather than on individual equilibriums, and despite Snow's patchy and self-directed medical training, he was peculiarly well placed to develop this.

First and foremost, he had a well-established interest in the physiology of respiration and the principles of gas exchange which he had developed during the early 1840s through studies on asphyxia and carbon dioxide poisoning.[66] He was familiar with the work of leading figures such as Magendie and the German chemist Justus von Liebig. He was well versed in the most up-to-date practical methods of chemistry and since his student days had pursued animal research as a means of investigating physiological questions.[67] Nor did the new French practice of using technology in clinical practice daunt him: he had developed a pump to administer artificial respiration to newborn infants and designed a trocar and cannula for the aspiration of pleural effusions.[68] His knowledge and practical skills were underpinned by his conviction that medical

*Figure 2.3* Portrait of John Snow by Thomas Jones Barker. R G Snow.

progress was only possible when the laws and principles of chemistry and physiology informed practice. A secondary point was his particular understanding of pain. It is clear from many of his writings that, as an onlooker during his training years, Snow had found the sufferings of patients undergoing surgery to be truly awful. He was not alone in this and there are many accounts from his peers which convey the horror of surgery without pain-relief.[69] But Snow constructed the pain of surgical intervention as a direct risk – not only was pain purposeless, it increased the risks of surgery because of its effects on the body's physiology. This was to be the foundation of his claim that in the majority of circumstances, the risks of anaesthesia were less than the risks of surgery. The final point to be taken into consideration is his economic motivation. Snow was clearly ambitious. Yet after many years of hard work he had achieved little in economic rewards. By 1846, he had been working as a general practitioner for 10 years and was still struggling to make a decent living, relying on dispensary work and a teaching post at the Aldersgate School of Medicine, the future of which was uncertain.[70] From his point of view, ether could not have arrived in London at a more propitious time; it fulfilled his intellectual aspirations and presented an opportunity to attract new clinical work.

It was in Robinson's practice at the end of December 1846, that Snow had first seen a patient under ether. He was smitten by the 'state of perfect quietude' and 'entire absence of pain', but did not begin in practice himself.[71] Instead he undertook a range of chemical and physiological experiments to ascertain the properties of ether and then investigated its effects on animals. He began by examining the effect of temperature on vapour concentration and drew on the work done by the Manchester chemist, John Dalton, who had established the concept of saturated vapour pressure at the beginning of the nineteenth century. Dalton had published a table of comparative saturated vapour pressures of different liquids in 1808, one of which was ether.[72] Snow used this as the starting point for a series of experiments which he carried out using a eudiometer to measure the amount of ether in air at different temperatures. Once he had confirmed that temperature was the determinant of the concentration of the ether vapour, he designed an inhaler. He based it on one invented by Julius Jeffreys in 1836 to treat bronchitis with moist air. Made from tin, it was designed so that the air breathed by the patient had first passed over warm water.[73] Snow adapted the design so that the inhaler was placed in a bowl of water. In this way the temperature of the water could be easily adjusted and the strength of the ether vapour controlled (Figure 2.4). His recognition of temperature as the determinant of vapour strength indicates familiarity with Black's work on latent heat.

On 16 January 1847, Snow made his first address on the subject of ether to the Westminster Medical Society. He had been a member since his student days and despite his initial shyness and Yorkshire vowels, he had become a regular participant, willing to contribute to any debate. He explained to members that the effect of temperature on ether vapour had been 'overlooked in the construction and application of the instruments hitherto used' and was the reason why so many of them were ineffective in establishing insensibility. He described his prototype and explained how the inhaler would be

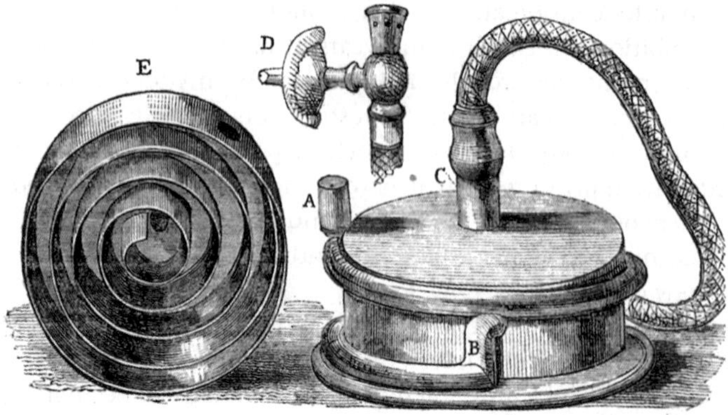


*Figure 2.4* Snow's first ether inhaler exhibited at the Westminster Medical Society on 23 January 1847. *L* I (1847) 121. Reproduced by courtesy of the Director and University Librarian, The John Rylands University Library, The University of Manchester.

metal – a good conductor of heat – and would use a waterbath to control the strength of the vapour. Economic aspects were clearly a consideration; he specifically noted that it would be both 'cheap and portable'.[74] A week later he returned to demonstrate his new inhaler to the Society. One of its most important features was a two-way tap which had been fitted to the breathing tube. This allowed the patient to 'begin by breathing unmedicated air, and have this gradually turned off as the etherised air is admitted in its place'.[75] The purpose was to try and minimise the initial excitement which was exacerbated by the risk of coughing and choking from the irritation of the ether vapour. His peers applauded his efforts, yet many found the explicit associations he drew between chemistry, physiology and practice were too radical to absorb. Hale Thompson, surgeon at the Westminster Hospital, castigated 'the mania which seemed to prevail of making experiments with improper or imperfect instruments', but most of his fellows, like the physician William Merriman, were more concerned to share bad experiences. Merriman recounted a case at St George's where the patient 'bawled out' as the knife touched him and he 'snatched his hand away'; and another where the inhalation was successful on the ward but when the patient came into the theatre 'the bandage fell from his eyes' and he 'seemed frightened'.[76] That ether had failed to work in such cases was apparent, but Merriman could not offer any explanation as to why.

Snow would not countenance that there could be such discrepancies when ether was administered correctly. As soon as he had established the principles of ether to his satisfaction, he looked for opportunities to practice and started to administer ether at St George's and University College hospitals. At University College, Liston swiftly regained his enthusiasm under Snow's efficient administration of the gas, and on completion of his first session at St George's, on 28 January 1847, Snow was thanked publicly by Caesar Hawkins, senior surgeon, who commented that the inhaler was far superior to ones they had previously used.[77] By 13 February, he had administered ether for eight surgical operations at St George's Hospital, and at this point Snow was able to say that in 'four-fifths of the cases' in which he had given ether 'there was not the least flinch or groan during the cutting by the surgeon's knife'.[78] It was only this type of case, where clinical observations and patient experience meshed, which could be claimed as a success. Even though the patient had no knowledge of pain, if he exhibited signs, then for Snow, the administration of ether was only 'partially successful'.[79] He continued to modify his apparatus and adjusted the size of the breathing tube, ensuring that it was neither so wide that it would choke the patient, or so narrow that air would creep round the edges.[80] Whilst building up practical experience, he developed his analysis of the process: drawing attention to the significance of the lack of oxygen in anaesthesia, as well as in asphyxia,[81] establishing the amount of carbonic acid gas produced during the inhalation of ether vapour,[82] and demonstrating the effects of ether on a green linnet at the Westminster Society, showing how the length of time the bird was left under its influence determined the outcome.[83]

By May, he had administered ether in over 70 per cent of the operations at St George's, as well as for many private operations; he was becoming established as an expert in its practice. He lectured medical officers from the United Services on the practical aspects of ether and demonstrated its effects on birds and small animals.[84] His advice was tailored to the military context – how a single-handed practitioner on board a small brig or schooner could employ ether, and suggestions on preparing for an inhalation in tropical climes. He was passionate about the way ether could ease the suffering of those who had fought for their country. He also remarked on its particular benefit in the services for discriminating between true and feigned illness, and recounted a case in which ether was given to a recruit who claimed injury to his spine, but when under its influence the spine relaxed and proved to be normal. Another case showed how a soldier's hip remained stiff and immovable under ether.[85] The removal of the patient's mind under ether gave doctors the power to exercise their authority over the body in a manner that had been impossible with mesmerism.

Snow's success in establishing ether may well have rested on his willingness to proceed with an inhalation regardless of the patient's response. If a patient became excited whilst breathing ether:

> talks, or sings, or laughs, or cries, and wants to move, ... he must be kept quiet, and the face-piece must be kept applied, even although he tries to get rid of it; for it would be wrong any longer to pay attention to his apparent desires, when he is not in a conscious and rational state. I have not met with any instance in which the patient could not be kept inhaling, with the assistance of somebody to hold his hand ... I prefer always to go on without stopping, and let the ether subdue the excitement it has produced.[86]

His confidence stemmed directly from his experimental work on animals and ether which had established the principle: 'that there is no person who cannot be rendered insensible by ether, ... [he] looked on the excitement as the occasional result of the cerebral functions being disturbed by a quantity of ether insufficient to suspend them altogether'.[87] Here we see the crucial difference between Snow and other doctors who regarded ether as a therapy and tended to reduce, rather than increase their administration of the gas if the patient became excitable.

Later that summer, Snow published a book on ether, to favourable reviews. He was described as 'experienced' and 'successful'; his 'ingenious apparatus' was praised and judged as 'equal in efficiency, if not superior' to other inhalers.[88] Despite the plethora of debates and accounts of ether in the medical journals and societies, Snow's was only the second full-length publication on the subject. Robinson had published a short volume in March, which contained brief details of his inhaler and surmised that 'the effect produced is probably that of intoxication, peculiar in its kind from the rapid manner in which it is induced; and the insensibility to pain which follows is analogous to that from ordinary drunkeness'.[89] In essence it was a collection of anecdotes rather than an authoritative guide: an example of biographical medicine. Snow's book was of a different mould; it was written specifically to address the practical issues. His administration of the vapour at St George's and University College hospitals had drawn a regular crowd of doctors and students who were keen to learn from his expertise; he had a natural bent for teaching. The book was intended to address the concerns of those trying to become adept in ether's use, but he found it impossible to discuss practice without first setting out the physiological background to the workings of ether. Here was the new relation between medicine and its 'collateral sciences'.

Snow had used the results of his chemical and animal experiments to analyse the process of anaesthesia into five identifiable degrees although he warned readers that:

> the division is, in some measure, arbitary – that the different degrees run gradually into each other, and are not always clearly to be distinguished, – and that the language I have used has been chosen with the sole object that my meaning might not be mistaken.[90]

Ether produced its anaesthetic effects through being absorbed into the circulating fluid and thus reached the nervous system. At this point it acted systematically and predictably, suspending function as it moved along the nerve pathways. The first degree of narcotism was the point when ether effected changes in the cerebrum. It altered intellectual functions but 'a person…retains a correct consciousness of where he is'. The second degree saw mental functions and voluntary actions being performed 'in a disordered manner' as the ether reached the cerebellum. At this point the patient resembled a drunkard or someone suffering from concussion. In the third degree, the effects of ether reached the spinal cord where it blocked sensitivity and motility; the mental faculties were completely suspended and the patient was perfectly secured against pain, even though 'muscular contractions' could occur. However, Snow recommended that in surgical cases, ether should be carried to the fourth degree where 'no movements are seen except those of respiration, and they are incapable of being influenced by external impressions'. The final effect of ether would be to act upon the medulla oblongata and cause death.[91]

The connections Snow was drawing between the external administration of ether vapour to the body and its sequential pathway through the nervous system were radical. He had taken the 'Yankee dodge' and placed it in the elite frameworks of anatomical and physiological knowledge, and then, through the use of analytical principles, he had tailored a specific intellectual framework for the etherisation process. But his approach differed markedly from most 1840s therapeutics which understood drugs to achieve their effects by diffusing through the whole system.[92] Snow was able to translate his findings into explicit practice guidelines because his experimental work had established the dosage of ether vapour needed to achieve each degree of narcotism. Thus he could state that an average middle-aged man needed to inhale air containing 45 per cent ether vapour for four minutes in order to reach the fourth degree of anaesthesia, which was suitable for surgery to commence.[93] Over later years, he found it better to reduce the concentration of the vapour to a 30 per cent mixture of ether and air and was able to anaesthetise children in two to three minutes and adults in four to five minutes. The length of time the patient had been inhaling was, according to him, one of the safest signs of unconsciousness. Length of inhalation corresponded to the amount of ether vapour taken into the system, although he recommended that if the administrator was in any doubt, he must use his clinical observations and wait until the 'excitomotory action of the eyelids diminishes, or till the breathing is decidedly automatic'.[94] There had been many instances where surgeons, acting on the 'appearance' of the patient who 'seemed' insensible, began to operate before the ether had induced complete insensibility. Thus, administering ether in quantifiable doses could, in Snow's view, produce an insensibility which for the patient predictably removed both the pain and the knowledge of an operation, and for the surgeon created a body that was as insensible as a corpse.

Snow was not the only doctor who sought to map out the process of etherisation: Francis Sibson, resident surgical officer at Nottingham General Hospital, had described ether's pathways through the nervous system, and other accounts of the 'phases' of the new technique were also published.[95] Wider afield, French and German physiologists explored the effects of ether. Flourens, building on his 1824 series of experiments, used trials on dogs to show how ether followed a predictable course: first affecting the intellectual functions through the cerebrum, then acting upon physical motor functions until it reached the spinal cord, at which point the patient became immobile and unconscious. He agreed with Snow that if ether was allowed to continue to the medulla oblongata, it would indeed produce death; and he also trialled several other narcotic agents including chloroform, which he dismissed as too powerful for human use.[96] Work by German physiologists Ernst von Bibra and Emil Harless analysed the chemical properties in etherised and normal animals. They found that in etherised animals, the fats of the brain and spinal cord were reduced but the fat content of the liver was increased. Their conclusion was that ether worked by drawing fats from the central nervous system and redistributing these to the liver, but they did not find similar patterns of redistribution of fats in blood or urine.[97] Their work was extended by clinical trials carried out by J. F. Heyfelder in Erlangen.[98] The novel technique was widely discussed at medical and scientific meetings, but there was no-one in Britain, Europe or America who combined the physiological skills with the practice of ether in the manner of Snow.

Snow was adamant (and was successful in convincing many of his London colleagues) that there was no danger to the patient if the ether vapour was administered correctly. But, during March 1847, the deaths of two patients whilst under ether had caused much alarm in medical circles. Ann Parkinson, the 23-year-old wife of a Grantham hairdresser, inhaled ether during an operation for the removal of a tumour on her thigh. She did not regain consciousness after the operation and died thirty-six hours later. At the inquest the coroner, Mr G. Kewney, stressed the seriousness of the case. If the death could not be attributed to either the condition or the surgery, then 'the person administering the ether' became 'answerable for the consequences', he warned. The verdict of the jury was:

> that the deceased... died from the effects of the vapour of ether, inhaled by her for the purpose of alleviating pain during the removal of a tumour from her left thigh, and not from the effect of the operation, or from any other cause.[99]

Robbs was cleared of any malpractice. In this instance, the legal process did not take account of any of the practicalities of the administration of ether, rather the emphasis was upon whether or not the clinical decision to administer ether was appropriate. Robbs' exoneration showed clearly how, in the view of the coroner, his attempt to alleviate the pain of surgery was judged to have been appropriate. The event was reported in the *Times* and generated some correspondence. It also caused Roger Nunn, surgeon at the Essex and Colchester Hospital, to reconsider the circumstances surrounding the death of one of his patients.

In February, Nunn had performed lithotomy on Thomas Herbert, a 52-year-old man, using ether. Herbert never recovered and died several hours after the operation had been completed. There had been no suggestion that ether was the cause of death and no inquest. But on reading of the case of Mrs Parkinson, Nunn wrote to the *Lancet* giving details of Herbert's death and said he now believed that although ether had 'fulfilled its intended offices', it had also acted as a severe depressant on the nervous system, so much so that it was unable to recover. As a result, Nunn said, he was now inclined to look upon the pain of surgery as:

> a healthy indication and an essential concomitant with surgical operations, and that it is amply compensated for by the effects it produces on the system as the natural incentive to reparative action.[100]

The fatalities caused the *Medical Times* to ask if 'the great discovery of the age' should be abandoned? It claimed that 'numerous serious accidents of a less fatal character' had established a 'distrust and suspicion' of ether in England, and noted that ether had been banned in Zurich.[101] The *Lancet* spoke of talented surgeons who had 'steadily set their faces against the use of ether' because its effects were 'simply intoxication'.[102] By this point the initial widespread take-up of ether had diminished. Although ether remained in use for major operations in large medical centres like London and Edinburgh, in other communities – Wales, Aberdeen and Bristol for instance – its use had tailed off.[103] But explanations for its rejection focused on the practical difficulties of using ether, rather than on considerations of safety. William Keith, surgeon at the Royal Infirmary in Aberdeen, for example, explained that ether was abandoned because it failed to work satisfactorily.[104] Snow noted that some surgeons gave ether up as a failure because they found it impossible to induce insensibility without struggling or excitement.[105] It is possible to imagine that ether could have remained as a marginal technique, its use confined to communities where individuals had mastered its characteristics, had it not been for the discovery of a new anaesthetic agent in November 1847 by Scottish physician and obstetrician James Simpson.

## Rejuvenation

Simpson became the best-known advocate of obstetric anaesthesia and was to receive many honours for his efforts in promoting anaesthesia. His discovery of chloroform has been described many times, but is worth revisiting for the contrast it presents to Snow's carefully staged chemical and physiological explorations of ether. Like Snow, Simpson from his student days was concerned about the problem of pain. He had immediately adopted ether in his obstetric practice but rather than restricting it to 'difficult' labours – forceps deliveries for example, he was swift to employ it to alleviate the suffering of normal births. Although he had been highly successful in his adoption of ether, he recognised that its 'inconveniences and objections' – disagreeable smell, irritation and tendency to excite the patient – dissuaded many doctors from using the vapour with confidence. His belief that another chemical agent could replicate ether's physiological effects had encouraged him to test a range of volatile fluids in the hope of finding one that possessed its advantages without its disadvantages.[106] But rather than test chemicals on animals, Simpson's method was to carry out 'breathings' in the company of his two medical assistants, Dr Mathews Duncan and Dr Thomas Keith. Duncan later explained how:

> On the day of discovery I selected two or three from the collected [substances] . . . chloroform and other two or three as deserving more careful trial than I could make at the time. . . . At any rate having had considerable experience in all kinds of breathings I took particular notice of chloroform as the best and likely to be most useful judging from the effects on myself.[107]

That evening he drew Simpson's attention to chloroform. Together, Simpson, Duncan and Keith breathed the vapour and rapidly became unconscious. Simpson was the first to recover and recognised that his search for a new anaesthetic was over.

Chloroform had appeared in British pharmacopoeias since the late 1830s and was known for its narcotic and antispasmodic qualities. David Waldie, chemist to the Liverpool Apothecaries' Company, had drawn Simpson's attention to it,[108] and sometime during the summer of 1847, William Lawrence, surgeon at St Bartholomew's Hospital, had substituted ether in his surgical practice with a solution of chloroform in alcohol which he described as chloric ether. Lawrence found it less irritating for patients although it was considerably more expensive to prepare. It had also been tried by surgeons at the Middlesex Hospital but abandoned because of the cost.[109] Nevertheless, Simpson took full credit for the introduction of chloroform anaesthesia. His announcement on chloroform was made to the Medical and Chirurgical Society of Edinburgh on 10 November and the news spread swiftly, not least to Boston, the first home of ether. It produced, said Warren, the surgeon who had trialled ether with Morton, 'an excitement scarcely less than that of the discovery of the narcotic effects of ether'.[110]

In Britain, doctors were palpably relieved that another anaesthetic had been found. Chloroform certainly reinvigorated enthusiasm for the suspension of sensibility and the first meeting of the Westminster Medical Society following Simpson's announcement was devoted to the topic. Robert Greenhalgh, an advocate of pain-relief in his midwifery practice, summed up the benefits of chloroform in comparison to ether: 'more easily applied', produced 'no excitement', was 'more rapid in its action' and left none of the 'unpleasant sensations' which followed ether.[111] There was nothing but praise for the effects of the new anaesthetic. The excitement caused by ether had raised issues of propriety because of the manner in which it appeared to place patients outside social or medical norms of control. The *Medical Times* had questioned the cause of 'those erotic feelings which have been noticed, especially in females, and convulsive phenomena observed occasionally in both sexes under the influence of ether'.[112] Chloroform induced unconsciousness quickly and effectively, thereby securing medical authority over the body; there were no reports of doctors failing to establish insensibility.

Snow immediately commenced chemical experiments to ascertain the ratio between chloroform and air, and animal experiments to determine dosage. He also inhaled chloroform himself until he became unconscious and was afterwards 'very sick'. Only eight days after Simpson had publicised chloroform, Snow used it during an amputation of a breast at St George's Hospital; he declared it to be 'equal to the best cases of etherization'. Nevertheless, his experiments had shown that the gas acted far more rapidly than ether and he warned members at the Westminster Society meeting that 'greater care was required in its use to avoid accident', precisely because of this quicker action.[113] It would only be a matter of weeks before this prescience came true and the first death under chloroform occurred.

## Hannah Greener – the first fatality

On 28 January 1848, doctors attended 15-year-old Hannah Greener at her home in Newcastle; she was to have a toenail removed. It was not the girl's first experience of anaesthesia; she had previously had another toenail removed using ether in Newcastle Infirmary. This time she appeared to 'dread' the operation and 'fretted a good deal'. However, she sat in a chair and began to inhale chloroform from a handkerchief. Mr Meggison, the surgeon giving the chloroform, explained how he tested her insensibility by lifting her arm which was 'rigid' and then directed Mr Lloyd to begin the operation. As the incision was made, Hannah gave 'a kick or twitch' and Meggison, assuming the movement indicated pain, was about to give more chloroform when Hannah's lips 'became suddenly blanched' and 'she spluttered at the mouth as if in epilepsy'. The surgeons dashed cold water in her face, gave her brandy, and then laid her on the floor to open a vein in her arm and the jugular vein. All to no avail and Meggison later reported that the 'whole process of inhalation, operation, venesection and death, could not have occupied more than two minutes'.[114] Following Hannah's death, a post-mortem was carried out by Sir John Fife, surgeon at Newcastle Infirmary, and Mortimer Glover, lecturer in Materia Medica at Newcastle Medical School, who had written a prize-winning essay on the physiological effects of chloroform in 1842.[115] During the autopsy, Fife and Glover found Hannah's lungs to be congested, though not collapsed, the stomach distended with food and the heart and the great vessels 'quite healthy'. Fife's evidence to the inquest placed the cause of death upon the 'congestion of the lungs' – a state of asphyxia – and he explained that this had been caused by 'some peculiarity in her [Hannah's] constitution – not to be detected beforehand – either in the lungs or in the nervous system'.[116] As in the case of the Grantham ether fatality, it was a verdict which exonerated the doctors from any responsibility for the effects of the anaesthetic.

This first death under chloroform received wide publicity. Simpson agreed with Fife's interpretation of the cause of death as asphyxia but attributed it to the brandy and water which had caused Hannah to choke, rather than the vapour.[117] Snow entered the debate but suggested that Fife had misread the details of the cause of death. He believed that death had occurred from overdosage of chloroform which had poisoned the heart, and he specifically related this to the method of its administration – a handkerchief. He explained that the rigidity of Hannah's arm after she had inhaled for about 30 seconds indicated that she was in the third degree of anaesthesia. Even though the cloth was removed at that instant, he suggested the effects of the vapour could increase 'at the same pace for twenty seconds longer; and at the end of fifty seconds from the commencement she would be in the fifth degree of narcotism' – the point of death.[118] Sibson, who like Snow had developed a special interest in ether practice, also protested that Fife had misinterpreted the facts and argued that: 'the immediate cause of the instantaneous death lay in the heart. The heart, influenced by the poison, ceased to contract, not from the cessation of respiration, for the heart in asphyxia will beat from one to three minutes after respiration has ceased, but from immediate death of the heart'.[119] But for Snow, the inquest verdict had wider implications than the technical explanation of death. His concern was that Fife's reading depicted the new technique to be unpredictable and difficult to control. This would, he noted:

> necessarily invest [anaesthesia] with some degree of danger, however small, and would entail some anxiety on both the operator and the patient. My view of the matter, holds out more hope for the future. I look on the result as only what was to be apprehended from the over-rapid action of chloroform when administered on a handkerchief...and consider that danger may be avoided by adopting another method.[120]

The clash of opinion on this first death – had chloroform poisoned Hannah's heart or deprived her lungs of oxygen? – framed a dispute which was to dominate the practice of anaesthesia until well into the 1900s. It clearly showed the tensions between the different ways in which doctors understood the body.

## Chloroform reigns, UK

Over the course of 1848, six further deaths occurred from chloroform: one in Britain, two in America, two in France and one in Hyderabad.[121] The patients were healthy and due to undergo minor operations; several of them had breathed ether previously without ill-consequences. The deaths caused concern about the safety of chloroform, which had begun in Britain with Hannah Greener's death, to escalate worldwide. From this point, we see the beginnings of a marked divergence in the use of the two anaesthetic gases – ether and chloroform. In a few medical communities like Boston and Lyons, the occurrence of deaths under chloroform caused doctors to abandon the new anaesthetic and return to ether. Surgeons there believed that there was little need to continue using a dangerous agent like chloroform; ether was a safer alternative. But in Britain, as elsewhere, chloroform remained in use.

Snow, as we noted earlier, had warned of chloroform's greater risks within days of his first experiments with the gas. Like the majority, he switched his practice to chloroform on the grounds that 'an occasional risk never stands in the way of ready applicability'.[122] Nevertheless, his awareness of its risks prompted him on several occasions to remind doctors that ether was a far safer agent; those who did not wish to become skilled in managing the dangers of chloroform should use ether.[123] He stressed that the slower action of ether made it an ideal anaesthetic for students and practitioners without specialist knowledge. Unlike chloroform which could kill without warning, an overdose of ether produced asphyxial symptoms which gave practitioners plenty of time to rescue the patient. Snow emphasised that the only reason doctors chose chloroform was to avoid ether's 'stronger odour, more pungent flavour, and other little inconveniences'; both anaesthetics were equally efficacious in removing the pain and shock of an operation.[124]

The almost universal rejection of ether as a fallback to chloroform can be explained in part by the difficulties doctors had encountered in administering ether effectively. Many used inhalers to give the gas; we saw earlier how this often produced unsatisfactory results – excitement rather than insensibility – which failed to meet medical aspirations of patient control. Indeed, when John Bland Sutton, surgeon at the Norfolk and Norwich hospital, remembered early anaesthesia it was more akin to a rugby scrum than the peaceful quietude of Snow's administrations.[125] Chloroform, which was easy and swift to use, delighted those who had struggled with ether. Chitty Clendon, surgeon-dentist to the Westminster hospital, had found ether unpredictable but succeeded with chloroform from the start. In only eight cases out of a hundred, he reported, had he failed to remove pain.[126] For Clendon and others who wished to use pain-relief, chloroform appeared to be the only option. The communities where doctors did revert to ether were ones where successful methods of giving the gas had been established well before the introduction of chloroform.

In Boston, for example, the first trials of ether had been made using Morton's glass flask, but within weeks surgeons were forced to abandon their use of his inhaler because of the impending patenting procedure. As a result, it became common practice for ether to be liberally poured on to a sponge, which was then held closely over the patient's face whilst doctors held the patient down, to prevent any struggling.[127] This shift in technique meant that in large part, American practice avoided the initial excitement of ether; patients were literally doused in the vapour and became rapidly insensible.[128] It became such an effective method that even when Morton dropped his patent claims, doctors did not return to using inhalers. In Lyons too, the surgeon Petrequin had pioneered a similarly forceful approach. He noted the general advice that 'the administration of ether should be graduated, intermittent, limited' was defective, and 'likely to prolong its operation, to exaggerate its inconveniences, and give rise to intoxication'.[129] But whereas surgeons in Boston and Lyons were comfortable with this 'heroic' method of ether administration, in Britain, doctors were reluctant to use force in their relations with patients.

We saw in Chapter 1 that British surgical culture in the 1840s was shaped by social and medical expectations of control; patients were expected to display fortitude and self-control, in addition to being tied down or held by assistants. This surgical anticipation of patient self-control explains why some surgeons found ether so difficult to establish. In many instances ether removed the self-control of patients to an extent which appeared to exacerbate, rather than ease the surgeon's requirements for successful operations. The way in which surgeons sought to establish patient cooperation with the process of inhaling ether suggests that the relationship between surgeon and patient was based on mutual dependency rather than surgical autocracy.[130] Patients were often offered a 'trial run' of ether to familiarise them with the process; if they expressed a dislike of ether during the first breaths, the administration would either be aborted, or resumed after much persuasion. In the operating theatre at St George's Hospital, in January 1847, surgeons found that 'all attempts, however to induce him [the patient] to inhale ether were fruitless'; the operation was abandoned. The next patient was 'exorted...earnestly to inspire the ether' by the surgeon, Henry Johnson, with eventual success.[131] Thus within this context, the choice of doctors to tolerate the risks of chloroform, rather than return to the difficulties of ether becomes meaningful. We shall see in later chapters how, despite regular fatalities, chloroform remained the dominant anaesthetic in Britain throughout the nineteenth century. One of the consequences of the persistent employment of chloroform was that disputes on the mode of anaesthetic death remained at the forefront of medical debates.

## Guises of death

From the time of the first ether trials in Britain, some surgeons had interpreted the process of inhaling ether as one which created a state of asphyxia in the body by depriving the blood of oxygen.[132] As we saw in Chapter 1, asphyxia had become a staging post in the process of death, from which patients could be recovered through artificial respiration. It was well recognised that asphyxia, frequently referred to as a 'handmaiden of death', created the loss of sensibility and it was because of the analogy between this state and ether that some doctors adopted the practice of 'resuscitating' patients from an inhalation by administering a few whiffs of oxygen gas.[133] When Snow began to investigate ether, he too considered that its action might be partly accounted for by the way the vapour displaced 'a great deal of the oxygen of the air' and caused a 'kind of asphyxia'. But when he tested this proposition through experimental work on mice, he found that supplying oxygen gas did not counteract the effects of ether. 'Asphyxia was a very different state from that produced by ether', he observed and speculated that ether worked by reducing to a minimum the 'oxidation of nervous and other tissues'.[134]

He continued his analysis by examining the out-breath from animals under ether and found that the vapour was breathed out by the lungs 'unchanged'; he noted that the amount of carbonic acid gas produced during the inhalation of ether was actually less than at other times; and he established that when the vapour of ether was mixed with air, it could prevent the oxidisation of phosphorous.[135] All these findings, he believed, were supportive of his earlier speculation. Flourens, who performed parallel experiments in France, agreed with Snow that the state of asphyxia was one where 'the nervous system becomes paralysed through the action of... blood *deprived of oxygen*' whereas with ether:

> the nervous system became paralysed primarily through the direct action upon it of this singular agent... this death by successive stages in the nervous system is the real point and the great point brought out by the new experiments... This isolation of *life*, of the point, the *vital knot (noeud vital)* in the nervous system, is certainly the most striking feature. In the etherized animal a single spot survives, and so long as it survives all the other living parts have at least a *latent life* and can resume full life: that single spot being dead, all dies.[136]

Flourens' colleague, Longet, however, noted that 'death from overdosage appeared to be due to a kind of asphyxia' which was directly linked to the effects of ether on the medulla oblongata.[137] Because few fatalities had been directly associated with ether, explanations of its mode of death remained of limited interest until the switch to chloroform and a rapid succession of deaths like Hannah Greener's changed matters.

Fife's assessment of Hannah Greener's death added support to the view of anaesthesia as a process of asphyxia, but Snow and Sibson suggested that the matter was more complex than this simple analogy. Sibson gathered information on each of the first four chloroform fatalities and concluded that in every case ' the immediate cause of the instantaneous death lay in the heart'. He suggested that there were two types of anaesthetic death: cases where the respiration ceased before the pulse and where life could be restored through artificial respiration, and cases which were 'almost hopeless', in which the first indication was the cessation of the heart.[138] Sibson understood chloroform to work like a poison on the muscles in the heart; if the gas was allowed to extend its effects beyond the suspension of circulation and respiration for anaesthetic purposes, then death would follow.[139] Out of this understanding came Sibson's recommendation that the clinical context should be adapted to lessen the strain on the heart; the patient should lie down, for example, to minimise the effort needed to maintain the body's circulation. Yet it would seem that the danger of asphyxia remained real for Sibson: by January 1848 he had devised a chloroform inhaler which converted into an artificial respirator.[140]

Snow agreed with Sibson that chloroform could indeed kill in two ways. The first type of death – respiratory – was only seen in animals. In the clinical context there would be plenty of warning signs:

> If it were possible for a medical man to mistake or disregard the symptoms of approaching danger, and to go on exhibiting vapour of chloroform, diluted to a proper strength till the death of the patient, this event would take place slowly and gradually.... The action of the heart would survive the respiration; there would be a great tendency to spontaneous recovery, and the patient would be easily restored by artificial respiration.[141]

For Snow then, asphyxia was a process of death into which doctors could intervene to restore life; it could be readily controlled and the signs of danger were clearly visible. The second type of chloroform death – syncope – did not allow such control. Instead, death occurred suddenly, and without warning, as a result of an overdosage of chloroform on the heart. The process of death was completely internalised within the body, and invisible to the eye, even to that of an experienced doctor. In order to effect control over the process, the administrator needed accurate knowledge and the technical means of controlling dosage; for this reason, Snow focused his teaching on the use of an inhaler.

Nevertheless, Snow and Sibson's explanations won little support from those such as Patrick Black, chloroformist at St Bartholomew's, and Liverpool surgeon, Edward Bickersteth, both of whom published on the topic in the early 1850s.[142] For Black and Bickersteth, the clinical characteristics of chloroform death were strongly analogous to asphyxia and it was this, rather than experimental work on animals, which determined their understandings of the process. But regardless of these differences over the mechanism of death, doctors agreed that quantifying the patient's experience of insensibility was highly problematic.

## The language of insensibility

Although the physical state created through inhaling ether could be linked to a diverse set of practices – opiates, intoxication, poisoning, asphyxia – there was no easy single analogy for the patient's mental state during the period of full unconsciousness. The only recourse for doctors and onlookers was to draw on clinical observations and impressions to construct their interpretation of the patient's experience. The following example highlights the inherent difficulties of such an approach.

In January 1847, the surgeons treating Mary Ann Chambers, who had spent the previous 9 months in Queen's Hospital, Birmingham with an ulcerating foot, decided that amputation was the only option. Before the operation Mary Ann was given 'two experimental trials with ether' which 'to all appearance' produced insensibility. Surgery began and as Professor Knowles passed the knife through the skin 'a sudden but transient frown' passed over Mary Ann's face. As the knife cut through to the bone 'the patient uttered an audible moan, but it was evident to those who heard it that it was very different to the cry or shriek of an individual in a state of consciousness'. Mary Ann had not had her eyes bandaged before the operation, as was usual, and witnesses were adamant that 'she was in no degree apprehensive of what was passing around her'. Once her 'intelligence' returned, she asked if her foot had been removed but could not accept this was so until she had been lifted to see for herself her cut limb. She was asked if she had felt pain, or been aware of any sensation, but claimed not to have suffered in any way or to have any recollection of the experience. Those who had observed the operation mooted the question as to 'whether the knitting of the brows, and the moaning did not indicate suffering?' They concluded that her lack of knowledge of the surgical act was strong enough evidence of true insensibility to the pain of the operation.[143]

To understand their witness of anaesthesia, onlookers had to reconfigure their understandings of bodily sensibilities into a new form. Whereas on a sensible body, contorted faces, shrinking limbs and moans stood as expressions of suffering, on an etherised body such phenomena were erroneous and misleading. Those familiar with the new constructions of the nervous system could place their observations in this framework. Derby surgeon John Lindley noted that:

> when the knife makes its first plunge into the body, there is a twitching or tremor of the whole frame; this, however, is no more than might have been expected from what is technically termed the reflex actions of the nerves, and ought not to be regarded as any indication of suffering, although it is held with horror by non-medical spectators from the habit, no doubt, of connecting pain with every unusual contortion of the body.[144]

For Mary Ann, her lack of recollection meant that she too had to reconstruct her experience, using the physical evidence of her missing foot and the facts she gleaned from onlookers. For other patients, anaesthesia transformed the pain of surgery into a dreamworld. Patients' accounts of their experience of unconsciousness received wide publicity; the *Morning Chronicle* told of a 13-year-old girl who dreamt of the country, a timid youth who on gaining consciousness said he had 'been dead or something', and a dental instrument maker who had 'felt as if an evil spirit was endeavouring to triumph over him, but still his confidence in his own victory was predominant'.[145] Ether 'embellished and disguised' an operation, noted one editorial, adding that through its processes the agonies of surgery were reconfigured 'into poetry, and all without pain'.[146] But not all patients experienced the comfort of pleasant dreams, or of 'better', even 'celestial' worlds into which they had escaped from the pain of the knife. Sometimes, such 'other worlds' were more compatible with the nightmares experienced by opium addicts: 'The dream is of drowning; a gushing in the ears, a choking and a sense of being lost, without pain or struggle or effort to save one's self', reported one account.[147]

The language of these early accounts reveals the way in which the new and strange process of ether inhalation was unfamiliar to both patients and doctors. The emotional response of patients to operations was rarely noted in surgical records. Only occasionally was there a brief note on the patient's fortitude, or lack of it. Instead, the purpose of operation notes was to describe all aspects of the surgical procedure – type of incision, actions taken to prevent blood-loss and closure of the wound – and to situate the medical observations within the locus of pathological classification. The innovation of ether introduced a different language into these stylised reports. Patients struggled to express their experience; doctors and onlookers struggled to explain what they had seen. It was very similar to the response of Davy and his associates to the effects of nitrous oxide. Reports of these early ether trials show a strong contrast between the descriptions of clinical symptoms or the surgical procedure, which were made in medical terminology, and the patient's experience of ether which was built around the patient's own words in a manner which evoked the earlier 'patient's narrative'.[148] A 68-year-old man at the Middlesex hospital underwent lithotomy because the 'vesical tenesmus was incessant, amounting to extreme incontinence of urine'. Yet his experience of inhaling ether induced a 'dreamy and very comfortable state'.[149]

But, as ether became established the language changed. A new medical vocabulary emerged, largely derived from Snow's characterisation of the anaesthetic process as a series of degrees. In May 1847, Liston removed a tumour from the jaw of a young woman whilst she was insensible under ether. The record noted that as she inhaled ether 'she passed into the second degree and . . . sobbed and screamed very much', then 'she passed into the third degree . . . extended in a state of great rigidity . . . [she] became suddenly quiet, going into the fourth degree of etherisation about five minutes after the inhalation was commenced' and the operation began.[150]

It was a bare 6 months since Liston had first operated under ether, but during that time, anaesthesia had been ring-fenced as a medical practice which was experienced by patients but classified and recorded by doctors. Such a sea-change exemplified how firmly the Yankee dodge had become embedded within the medical domain and imbued with medical purpose. Breathing gases to enhance the intellectual sensibilities of life had metamorphosed into a process which enabled the self to transcend the suffering of surgery through the experience of 'near death'.

# 3
# Science Versus Empiricism

During the first 12 months of anaesthesia, two key advocates emerged and remained as figureheads of British practice throughout the nineteenth century. London practitioner John Snow and Edinburgh physician James Simpson were unanimous that pain-relief was justified in all cases of surgery and childbirth; they agreed that the patient should have a voice in the matter, and were equally condemning of those doctors who refused patients painless surgery for fear of the risks (Figure 3.1). But they diverged hugely when it came to the method of administration: Snow designed and employed inhalers, Simpson promoted the use of a sponge or handkerchief. Given their shared commitment to anaesthesia, this difference might seem insignificant. Certainly at the time, William Fergusson, Snow's lead surgeon at King's College hospital, dismissed the difference of methods as irrelevant.[1] Yet, when the reasoning behind each man's method is teased out, the divergence reveals far more than a simple practice distinction. Rather, it reveals how the new view of the body, as a universal system of tissues and organs, had reconfigured not just the knowledge and practice of medicine but the very identity of doctors. The sticking point between Snow and Simpson was science, not just in its broad meaning, which both affirmed as an irreplaceable source of new knowledge and understanding, but in the very specific way in which its bodies of knowledge – chemistry and physiology for instance – should feed into and shape medical practice.[2]

For Snow, experimental science was the anchor and mainstay of medicine; a doctor's responsibility was to mediate its possibilities to the public, using his special knowledge of its laws and principles to ensure safety. Simpson's universal use of anaesthesia was derived from a different set of values that had far more in common with the earlier patterns of Scottish Enlightenment medicine. In this chapter, I polarise

*Figure 3.1* James Young Simpson, frontispiece in Laing Gordon, 1897. Reproduced by courtesy of the Director and University Librarian, The John Rylands University Library, The University of Manchester.

the debate between the 'new' approach of Snow and the 'old' approach of Simpson in order to identify how and why the same medical process became established in two parallel methods of practice.[3] In that Snow's use of inhalers was followed by a large number of London doctors and supported by the medical journals, and Simpson's method (also promoted by Edinburgh surgeons, James Syme and James Miller) became known as the Scottish method, the difference was also one of national cultures and until the end of the nineteenth century it remained contentious and divisive. Both Snow and Simpson had died by the time of the Hyderabad Commission in the 1890s when the issue of anaesthetic method was again debated not just across the border but across the world.

## Snow and Simpson

Snow and Simpson were born a couple of years apart and shared many similarities: supportive families who believed in education, personal ambition, adroitness at using medical networks to promote their work and a readiness to adopt innovative practices.[4] Their medical training spanned the spectrum of options in the first part of the nineteenth century. Simpson graduated from Edinburgh University in 1830 and after assisting in general practice in Falkirk and Bathgate (his home town) returned to complete an Edinburgh MD thesis on death from inflammation; by 1835 he was assistant to the professor of pathology. Snow, as we saw, entered medicine through an apprenticeship and a self-made package of clinical training in London, at the Hunterian School of Medicine and Westminster Hospital. By 1838 he had become a Licentiate of the Society of Apothecaries and a Member of the Royal College of Surgeons. This was no mean feat for a working-class man without patronage or financial resources. He knew that to succeed as he wished he needed the status of university qualifications and he sat the MB of London University in 1843, followed by the MD in 1844. Simpson meanwhile had become Professor of Midwifery in Edinburgh in 1840, whilst only 29-years old. Both men made full use of the professional networks of medical societies and publications; neither shirked controversy with their peers. In London, Snow regularly addressed meetings of the Westminster and Royal Medical and Chirugical Societies; in 1838 he challenged the opinion of E. F. Lonsdale, anatomy demonstrator at the Middlesex Hospital, on the structure of the recti muscles and he published on topics such as asphyxia and stillborn infants (1841) and carbon dioxide poisoning (1839, 1844, 1846).[5] Simpson revelled in debate, never refraining from using rhetorical force to castigate opponents. Between 1845 and 1847 there was a particularly virulent disagreement between Simpson and Robert Lee, lecturer in midwifery at St George's, on the correct treatment for placenta praevia.[6] 'We shall see', insisted Lee at the tail of the debate, 'whether Dr Simpson himself does not flee off to some new marvel, some fresh novelty to attract public notoriety, and to cover his defeat in the BATTLE OF PLACENTA PRAEVIA'. Lee's prophecy appeared in the *Lancet* adjacent to Simpson's announcement of chloroform which was headed, *On a new anaesthetic agent, more efficient than sulphuric ether*.[7] The irony could not have been bettered in fiction and needless to say, Lee was soon strongly opposed to the use of anaesthesia in childbirth.[8]

By 1846 then, Snow and Simpson were established within their different networks and actively engaged with some of the pressing questions of the time. The most striking difference between them was in the success of their clinical practice. Simpson, by this time married and with a growing family, was a notoriously popular physician and obstetrician, enjoying a thriving private practice among the elite of Edinburgh and London. Snow ran a small general practice from his lodgings in London's Soho, an area dominated by the working classes and with a swift turnover of residents. Around 60 patients are listed on his books for 1848; he eked out his income with dispensary work and lecturing. Both men were quick off the mark to learn more about the Boston ether trials: Snow visited Robinson, the dentist, to see him extract a tooth under ether on 28 December; and between Christmas and New Year, Simpson travelled to London to visit his old university lecturer, Liston, and learn of ether first-hand.[9]

The possibilities of preventing surgical pain by inhaling ether vapour enthused both Snow and Simpson, and the divergence of approach is discernible from this very early point. Before January was over, as we noted in Chapter 2, Snow had obtained ether, subjected it to a range of chemical analyses, designed an inhaler and begun a series of physiological experiments on animals. He had also started practice at St George's and University College hospitals. Simpson, notwithstanding his concerns about its possible adverse effects, had used ether during a labour in which the outcome was doomed by the mother's contracted pelvis; her previous labour had lasted four days and the dead infant had been extracted using instruments. Although ether did not save the baby, it did relieve the mother's pain and she spoke 'with gratitude and wonderment of her delivery'. Simpson too was euphoric: 'I can think of naught else' he wrote to his brother, and claimed it to be a far more significant event than his appointment as Queen Victoria's accoucheur in Scotland which was made on the same day.[10]

The distinguishing feature between the two men's perception of ether was that Snow, from the very beginning, saw the innovation as a 'new branch of science', separate and distinct from any existing medical practice and one which was firmly grounded in chemistry.[11] One of the best examples of his staged development of principles was a series of 18 articles, written for the *London Medical Gazette* between May 1848 and December 1851 – *On Narcotism by the Inhalation of Vapours*.[12] These covered his physiological research into ether and chloroform; they recorded his search for a superior agent, one which would combine the efficacy of chloroform with the safety of ether; and they contained a wealth of clinical guidance.

The serialisation of this experimental work encapsulates the very sense in which it was knowledge in the making. At one point he apologises to readers for the delay between articles. This was occasioned, he explained, by the desirability of repeating 'many experiments and institut[ing] fresh ones, the performance of which occupied a great deal of time'.[13] His clear, direct style conveyed how these early researches involved a learning curve as he gained experience in exploiting the possible variables within an experiment. So whilst recounting how a succession of mice, guinea pigs and birds were put into jars containing different strengths of chloroform vapour (Figure 3.2) he noted that a thermometer was used to take the temperature of the guinea pig and linnet at the end of their exposure to the vapour but not in the other experiments in the series because, 'these are the only occasions on which it occurred to me to apply the thermometer'.[14] Often the next group of experiments grew out of his observations of an earlier set. So for example, having determined that one grain of chloroform in each 100 cubic inches of air was enough to induce the second degree of narcotism, it 'occurred to me' he said, 'that if this method of ascertaining the amount of vapour in the blood were correct, then a much more dilute vapour ought to suffice to produce insensibility in animals of cold blood; and that experimenting on them would completely confirm or invalidate these views'. He demonstrated the point in a lecture at the Royal College of Physicians by placing a chaffinch and a frog in the same jar of chloroform, whereupon the very dilute vapour produced insensibility in the frog, whilst the chaffinch remained unaffected.[15]

Experimental work confirmed the findings of other investigators: a report by the Frenchmen Dumeril and Demarquay that animal temperature dropped during inhalation of ether and chloroform was proved on a green linnet – the temperature fell from 110° to 102 °F between the beginning and end of a period in a jar of chloroform.[16] The London medical community were aware of his research and many were supportive of his quest; the chemist Mr Bullock and Robert Barnes, obstetric physician at the London Hospital, supplied him with samples of nitric ether which he proceeded to test on mice; having established its 'sparing

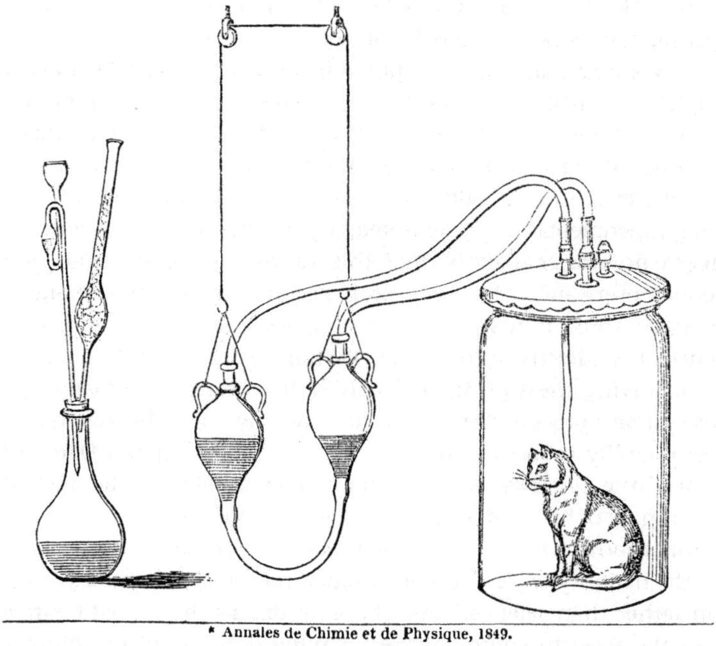


*Figure 3.2* Snow's experiment on the excretion of carbonic acid by animals whilst under the influence of chloroform and ether taken from 'On Narcotism'. *LMG* 12 (1851) 622. Wellcome Library, London.

volatility' which produced anaesthesia gradually, he gave it to a patient at St George's for a tooth extraction.[17]

The ease with which Snow drew on chemistry and physiology to determine the principles of anaesthesia was singular but not unique. In 1830, whilst Snow was an apprentice in Newcastle upon Tyne, John Herschel published his *Preliminary Discourse on the Study of Natural Philosophy*.[18] Herschel, son of the astronomer William Herschel who classified stars,[19] was clear that the focus of scientific enquiry should be 'the operation of general causes, and the exemplification of general laws' and claimed that 'every object which falls in his [the scientist's] way elucidates some principle, affords some instruction' which in turn leads to 'a sense of harmony and order'.[20] 'We must never forget', he says, 'that it is principles, not phenomena, – the interpretation, not the mere knowledge of facts, – which are the object of enquiry'.[21] Snow's use of Herschel's two-pronged method of observation and experiment – induction and deduction – is apparent from his earliest experimental work.[22] So too is his adoption of Herschel's inclusivity: 'there is scarcely any natural phenomenon which can be fully and completely explained in all its circumstances, without a union of several, perhaps of all, the sciences' observed Herschel and explained how establishing a law in one branch of science 'immediately furnishes us with a means of extending our knowledge of innumerable others'.[23] If during experimental work, he noted, a point was reached at which 'we perceive no further analysis', then this was the cue for 'the study of that phenomenon and... its laws [to] become a separate branch of science'.[24] Thus Snow's perception of anaesthesia as a 'new branch of science' fits directly into Herschel's framework, as indeed does his accompanying stress on the difficulties of creating a new language to express a new process.[25] Snow was not the only one of his generation to be profoundly influenced by Herschel's methods. Charles Darwin too had read the *Discourse* as a student and drew heavily on the method in the planning of his arguments for the *Origin of Species*.[26]

Snow's most notable flair, though, was his perspicacity in realising that the majority of his peers determined their practice through empiricism rather than science.[27] And because of this, he thought carefully about the need to communicate his understanding of the process of anaesthesia in a manner which could be understood and adopted by all doctors. The sharing of experimental findings was taken by Snow to be a primary responsibility of his position; he engaged widely through the networks of medical societies, publications and journal correspondence. (He was to follow the same pathways to promote his theory that cholera was a water-borne disease from 1848 onwards.[28]) But he knew that in order to achieve a take-up of his findings, he had to shape and package the information in a form that would be meaningful to his peers. Some mediation was needed between bench and bedside. Whereas in experiments 'the most perfect way of giving a vapour to animals' was by immersing them in jars of chloroform or ether vapour so the breathing was not interfered with and 'the strength of the vapour [was] accurately known', in a clinical context 'our endeavour should be to approach it as nearly as we conveniently can'.[29] It was this ability to shape the findings of experimental science into clinical principles which underpinned Snow's approach.

We saw earlier how his description of the process of anaesthesia, as a set of sequential and progressive degrees of narcotism, accorded with the findings of the French physiologist, Flourens. But the way in which he distinguished his explanation from that of Flourens' revealed his appreciation of the focus of London practice: 'The division I have made from observations on patients', he said, 'will... be found better for practical purposes... it involves no theory about the functions of the nervous centres, which is perhaps an advantage, as those particularly of the cerebellum, are probably not definitively known'.[30] In London, most doctors drew on empirically grounded knowledge. 'I have often heard a medical man say', Snow noted,

> on lifting a patient's arm and seeing it drop down again, "the muscles are quite relaxed, he is under the influence of ether now", at a time when the state of the muscles merely depended on volition not being exerted on them, and when a cut would, undoubtedly, have roused a vigorous resistance.[31]

For Snow 'mere observation' was not enough; 'knowing' that a patient was insensible enough for surgery to begin consisted of a combination of objective facts – length of inhalation and strength of vapour – supported by clinical observations of the state of the patient and the ability to 'place' these within the process of anaesthesia. In his view, the facts could be derived from the use of inhaler – it was a tangible embodiment of the principles of science – and when these were coupled with knowledge of the effects of gases upon the physiology of the nervous system, it made the administration of chloroform 'unattended with danger'.[32] Snow drew his authority to initiate a process of death in a patient from this specialised body of knowledge which had its origins in chemistry and physiology, rather than in humanitarianism.

For Snow, the purpose of technology was to support and aid practice; he was intent on ensuring that the design of inhalers should mesh with the clinical context. His ether inhaler was designed with a chamber for the chemical which was deep enough to withstand the 'vigorous and deep inspirations' of the patients, and prevent the ether being 'agitated' and 'splashed into the elastic tube'. The tube was around three foot in length in order to leave as much room as possible for the surgeon and assistants. The valves which controlled the admission of air were 'light' and positioned so that they worked 'in any posture in which the patient can be required to be placed'. His face-piece was lined with oil-silk and had been altered to accommodate 'faces of different dimensions'; (Figure 3.3) he subsequently produced a face-piece specifically for children.[33] Later his chloroform inhaler followed a similar pattern – blending chemical laws into clinical context.

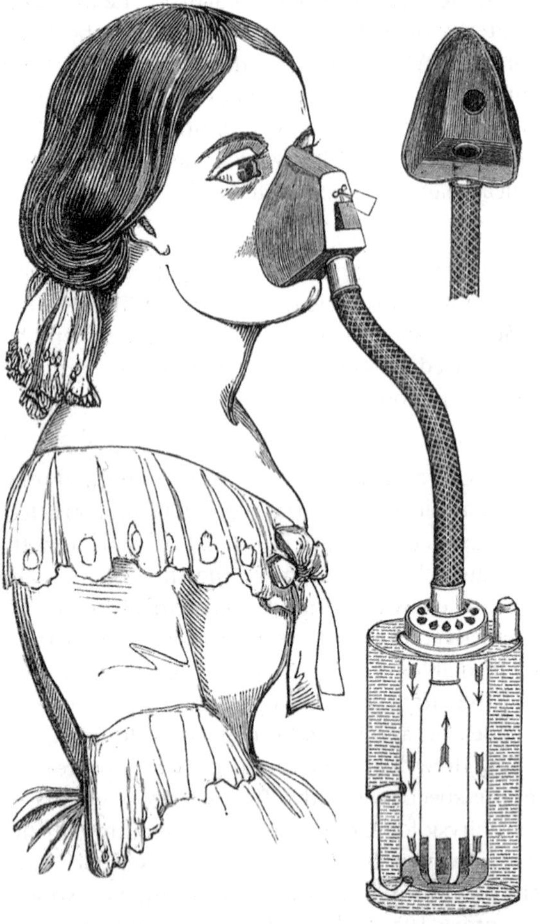

*Figure 3.3* A perfectly controlled Victorian lady demonstrates Snow's chloroform inhaler and mouthpiece. Taken from *OC*. Reproduced by courtesy of the Director and University Librarian, The John Rylands University Library, The University of Manchester.

In Snow's view, the other method in vogue, 'the handkerchief', which was advocated on account of its 'supposed simplicity', was instead 'a very complicated process, on account of the difficulty of getting even an approximative knowledge of what I was doing, by the best calculation I could make'.[34] What appeared to be a utilitarian artefact with public appeal became complex and problematic in the context of the new scientific medicine.

## Inhaler or handkerchief?

The debate over inhaler versus handkerchief had begun in February 1847, when, at the time that Snow was refining his first apparatus, Thomas Smith of Cheltenham told of a 'most simple contrivance for the effectual and safe administration of ether'. This was a sponge saturated with ether, and applied over the patient's nose and mouth.[35] Samuel John Tracy at St Bartholomew's Hospital followed Smith's example and north of the border so did Simpson, Syme and Miller in Edinburgh.[36] By the time Simpson published his news of chloroform, Smith's method had become widespread throughout Scotland. From the first, the sponge was associated with clinical efficacy.[37] The failures attributed to ether resulted from the method, asserted Simpson, and use of a sponge avoided the primary state of exhilaration 'by impregnating the respired air as fully as the patient can bear with the ether vapour and allowing it to pass into the lungs both by the mouth and nostrils so as rapidly to complete the anaesthetic effect'.[38] It was indeed very similar to the method adopted by Boston surgeons, discussed in Chapter 2, and avoided the difficulties of excitement and struggling created by the more tentative administrations of the gas through inhalers. By November 1847, Simpson was extolling the virtues of a 'simple hanky' (Figure 3.4). It was, he continued:

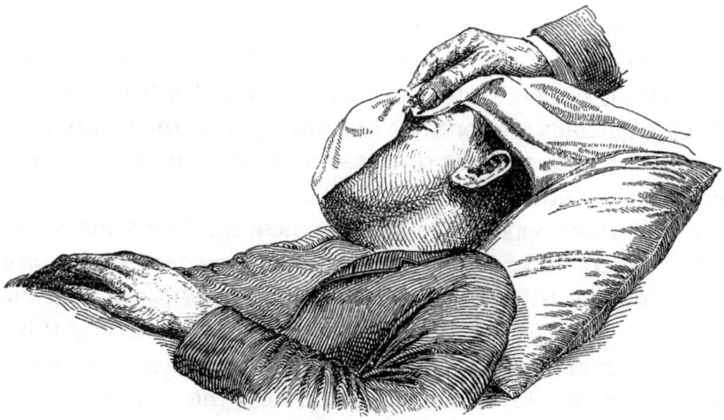

*Figure 3.4* Simpson's patient-friendly method of administering chloroform on a towel. Taken from Hewitt 1893, p. 191. Reproduced by courtesy of the Director and University Librarian, The John Rylands University Library, The University of Manchester.

> far preferable to every means yet adopted... infinitely preferable to any instrument yet seen, some of which exhibit it by the mouth and not by the nostrils, in small and imperfect, instead of full and complete doses; and with instruments so constructed there is no doubt whatever that failures and exciting events would ever and anon occur.

Besides, he avowed, 'inhaling instruments frighten patients, whilst the handkerchief does not'.[39]

His Scottish peers James Syme and James Miller supported Simpson's views. Miller was equally condemning of inhalers: 'Many machines have been invented. I believe they are all useless and not a few decidedly mischievous.' He waxed lyrical, however, about the range and accessibility of the simple means which could be used to bring chloroform to the mouth and nose: 'a handkerchief, a towel, a piece of lint, a worsted glove, a nightcap, a sponge... In the winter season, the glove of a clerk, dresser, or onlooker, has been not unfrequently pressed into service' he declared.[40] (In private practice of course, the most common choice was a 'pocket silk handkerchief'.) Simpson believed that giving ether or chloroform was comparable to other therapies, which were judged by their 'effects more than quantity'; his aim was for the patient to become 'thoroughly and indubitably soporized' by the vapour.[41] Miller extended this analogy comparing it to remedial bleeding. 'We do not think it necessary', he pronounced, 'to be telling of the ounces as they flow, but are regulated entirely as to the amount drawn by the effects produced, so we as little think of dropping, or otherwise of measuring the chloroform'. In Edinburgh, the objective was to 'produce insensibility as completely and as soon as we can; and there is no saying, *a priori* whether this is to be accomplished by fifty drops or five hundred. We begin with generally two or three drachms spilt on the handkerchief or lint; and we refresh that, or not, from time to time, as circumstances require', he concluded.[42]

Snow took every opportunity to reject these arguments and to explain that the method was essentially flawed; at the very least the chloroform should be diluted with spirit to lessen the danger.[43] Provoking though the Scottish claims were in his quest to educate his peers on this 'new branch of science', his tone during these interchanges remained calm. Only occasionally did irritation sear through the surface such as the footnote in *On Chloroform* which referred to Miller's 'somewhat slovenly, and not very cleanly manner'.[44]

Entertaining though such disputes may be, they are more than vignettes of professional rivalries and conflict; the differences between the inhaler and the handkerchief reveal much about mid-nineteenth-century constructions of the body.

## Patient bodies

By 1852, Snow believed he had quantified the dangers of the Scottish method. In a paper on *The cause and prevention of death from chloroform* read to the Medical Society of London, he told how he had collected details of 18 chloroform fatalities. In 16 of these, the chloroform had been given on a hanky, towel or piece of lint and in the other two cases, an inhaler was used but not 'by a medical man'. A wide-ranging discussion followed. Members spoke of the 'peculiar effects' of chloroform which could affect the constitution longterm, the difficulty of obtaining pure samples, the responses of 'hard drinkers' to inhalation. Few picked up on the connection Snow had drawn between method and mortality. George Barlow, physician at Guy's, the only member to specifically comment, said dismissively that 'the danger of administering chloroform on a handkerchief had been much overrated: it might be done safely if there be not too much on it'.[45] A few weeks later, Snow wrote to the *Medical Times and Gazette* and explained that following the publication of his paper he had been contacted about a chloroform fatality in Glasgow. The absence of coroner's inquests in Scotland meant that there was no necessity for practitioners to report deaths and he suggested that Simpson, with his 'influence and connexion', would be able to obtain and publish data on Scottish chloroform fatalities.[46] Simpson duly replied. But he gave no new information, only a general claim that the two cases mentioned were the only fatalities 'among the many thousand cases' of surgery in which chloroform had been used.[47]

Early in 1855, Syme heightened the stakes of the debate by suggesting that Scottish chloroform practice was far safer than that which was practised south of the border. It was a time when London doctors were particularly vulnerable to such charges. As we shall see in Chapter 4, a cluster of fatalities in London practice during 1853–54 created a crisis of confidence in anaesthesia and caused surgeons to sharply curtail their use of chloroform. London doctors invested anaesthesia with intense risk, stated Syme, whereas in Edinburgh, with moderate care, it was deemed 'perfectly safe'. In his operating theatre, where it was given 'almost daily', there had been no fatalities.[48] The difference between London and Scotland, he said, was attributable to one of three things: the chloroform, the patients or the method. Most chloroform used in London was manufactured in Scotland.[49] And Syme went into great detail to prove that London practitioners were far more choosy about the patients who underwent anaesthesia, whereas we, he said nonchalantly, 'never ask any questions as to the state of the heart'. He concluded that the difference in results depended on the alternative method of procedure and drew a very pertinent distinction between the two places. In Scotland, said Syme, 'it is given according to principle, there [London] according to rule' and he proudly underlined the Scottish approach as one which was 'guided only by the symptoms of the patient'.[50]

Snow was so roused by reading Syme's claims that he immediately wrote a long response in defense of 'London as a whole'. He began courteously, supporting Syme in his view that heart disease was no bar to anaesthesia, but it was not long before he tackled the heart of the issue – Syme's claim that mortality was higher in London than Edinburgh. 'Mr Syme seems entirely to overlook the relative size of population of the two places. When these circumstances are taken into account, the mortality from this cause seems pretty equal' and in fact, said Snow, 'in nearly all cases where accidents have happened in London, the Edinburgh method of exhibiting chloroform was followed pretty closely'. 'It is my opinion', he continued, 'that the rules of some of us here proceed from the principles we have taken great pains to ascertain and establish'.[51] In his case, the rules had been derived from the findings of experimental science.

For those doctors who were following this dispute in the *Lancet*, Syme's distinction, between principle and rule, was a further example of the tensions created by the different models of the body. Syme's claim that the Scottish method placed the patient firmly at the centre of the proceedings drew on the biographical model of the body as an interdependent system. Disease was read as imbalance and the purpose of therapeutic intervention was to restore each body to a natural harmony. The constitution was a key determinant of physiological responses to intervention. As Miller had explained, although doctors worked within therapeutic principles, their skill was to manage treatment by reading the effects of therapy upon the body.[52] Ether and chloroform had been classified in the pharmacopoeias with poisons such as arsenic, mercury and so on. In therapeutic practice, doses of these were given to a patient until the desired level of purging or debilitation was reached; the essence of medical skill was to determine the point at which this was achieved. The Scottish method of giving chloroform, therefore, resonated comfortably within these paradigms. It gave recognition to the principle of anaesthesia – all bodies could be rendered unconscious and insensible – but no prediction could be made about dosage because each body journeyed into unconsciousness in its own individual and natural way.

'There is no uniformity of dose', proclaimed Miller.[53] Instead, successful anaesthesia depended upon the confidence of the administrator to persist in the administration of chloroform until they could see the required state had been reached.[54] They must keep their eyes pinned to the patient's face, advised Miller, and expect to see him 'snoring... his eyes fixed, his body pliant and motionless;... bluish in the face, sputtering saliva rather freely from the mouth and seeming to the inexperienced eye on the very verge of apoplexy'. This state of 'simulated apoplexy' was the required level of anaesthesia for surgery.[55] The simplicity of the method made it a perfect duty for medical students as it required no additional skills other than the acute clinical observation of the body during the inhalation. Because the danger of chloroform was understood to lie in asphyxia, the Scottish method was guided 'entirely by the respiration'. 'You never see anyone here with his finger on the pulse while chloroform is given.' Instead, said Miller, the administrator kept his eyes firmly on the patient's face, ready to open the mouth and 'seize the tip of the tongue with artery-forceps, and pull it well forward' if respiration became difficult or ceased.[56]

The method of giving anaesthesia through an inhaler, was drawn from a very different understanding of the body which as we saw in Chapter 1, emerged from the hospitals of post-Revolutionary France. It was a model which made possible the application of universal processes such as anaesthesia and supported Snow's claim that the benefits of insensibility were suitable for all ages and sexes of patient. There were indeed differences – children responded more quickly to anaesthesia because their circulation was quicker, alcoholics took longer to become unconscious – but these were only at surface level. Underneath the material trappings of clothes and skin, and the mental influences of culture, each body's nervous system would respond to ether and chloroform in a predictable and universal manner. This meant that anaesthesia could become a standardised practice: the amount and strength of each inhalation could be quantified because the parameters of safety were the same in all bodies. But for the doctor it was a process that demanded more than the clinical observation of *visible* changes to the body: instead, it required a knowledge about the *invisible* changes to the nervous system effected by the ether or chloroform. Knowing that the danger of chloroform lay in poisoning of the heart by overdosage, rather than asphyxia, meant that there was no purpose in focusing entirely on respiration. Clinical observations alone were of little value in this 'new' science. 'If a person who has not experimented carefully with chloroform on animals, nor seen a fatal case of its administration, were to judge entirely by his own observation', said Snow, 'he would probably conclude that danger began with the respiration (because it is in that function that a very slight overdose manifests its effects when the agent is well diluted with air), but he would be in error'.[57]

Thus the difference of opinion between Edinburgh and London as to whether the pulse or the respiration was the key clinical aspect of anaesthesia, arose because of the very different models of the body and its systems each method drew on. The distinctions also fed into Snow and Simpson's understandings and constructions of medical identity and it is to these that we now turn.

## Figures of authority

The best entrée to Snow's perception of medical identity can be found in his presidential address to the Medical Society of London in 1855 which expressed his vision that if all branches of medical practice could be established upon the analytical and systematic foundations of science then it: 'will be impossible that intelligent persons should submit to be treated with globules of sugar of milk, having the name of medicine attached to them and nothing else'.[58] In other words, patients would seek health care only from orthodox practitioners.[59] Snow believed that the only branch of medicine which had achieved such certainty was orthopaedics practice: 'all surgeons are agreed respecting the proper treatment of fractures and dislocations; they know what to expect from treatment, and if they are disappointed, they can generally explain the reason why'. And he read a direct payback into this newly established body of orthopaedic knowledge: 'the whole tribe of ignorant bone-setters have disappeared, or, at all events, they practise only amongst the most uneducated of the population'.[60] In his view then, not only did the use of sciences – particularly physiology because he regarded it as the foundation of pathology and therapeutics – endow medical knowledge with a unique authority, but it also formed the means of establishing successful relationships between doctors and patients by providing the boundaries which could govern expectations.

Nevertheless, the 'difficulty' of the sciences meant that much of the knowledge of the 1850s was 'imperfect'.[61] 'Practical medicine has advanced less rapidly than might have been expected from the progress of discovery in other branches of scientific research,' noted James Bird, president of the Harveian Society in 1853. The public will only accord us dignity and authority, he told his peers, if they believe we are applying 'skill in all the elementary sciences' to medical practice.[62]

The meanings invested in science by Snow and Bird and their perception of the urgency of the need to demonstrate 'advancement' in medical practice reflected more than intellectual concerns. The business of making a medical living in the mid-nineteenth century, particularly in London where elite physicians and surgeons dominated practice, was an intense and often unsatisfactory experience. Intellectual, economic and practical tensions were manifest at all levels, as has been well documented.[63] Thus Snow's belief that medical practice needed to be endowed with 'certainty' if it was to achieve social recognition and status was not only of esoterical value. Rather it translated into the gaining of public confidence by establishing transparency in patient expectations of treatment outcomes. This in turn would boost the status of doctors in the public eye and ensure that 'the civil engineer and the chemist' would not be 'placed over the medical man in matters which belong exclusively to his own profession'.[64] Science would bring authority and status to medicine, as it had for public health officers who were emerging as experts on the health of the nation.[65]

Some historians have attributed the sense of tension in patient–doctor relationships in the mid-nineteenth century to sociological factors such as an overcrowded marketplace and indeed, that was a significant contributory factor.[66] But one of the greatest sources of tension in patient–doctor relations during the nineteenth century rested on the difficulties doctors faced in marrying the objectives of the new scientific medicine with the established traditions of biographical medicine.[67] The new scientific medicine removed commonality between patient and doctor because it drew on a body of medical knowledge which was closed to the patient. The greater number of patients knew 'little of medical science' and this placed great responsibilities upon doctors, observed Snow.[68] The doctor's role was to mediate the benefits of medical science to the public; he was the gatekeeper between this closed body of knowledge and the public domain. But privilege begat responsibility and the doctor needed to use his authority to protect patients from being 'duped' by 'unscientific' practices or exposed to unnecessary risk. It was an active role, demanding full engagement with the most recent advances of medical sciences; no doctor should administer therapies without some understanding of their effects, cautioned Snow.[69] There was a fine balance in ensuring that an allegiance to the principles and laws of physiology and chemistry was not overtaken by a doctor's moral desire to prevent suffering and prolong life. Snow accepted that this balance did not come easily to all doctors. But in the case of anaesthesia, both objectives could be met through doctors becoming fully informed about the physiological effects of chloroform and using an inhaler to prevent overdosage, or by returning to ether, which was a far safer agent.[70] This way, a doctor gained the humanitarian freedom to remove pain without compromising scientific principles. That was the role that Snow was working to create.

Turning to Simpson, it is possible to sketch out a medical identity which was far more characteristic of biographical medicine and carried particular resonances of Scottish Enlightenment philosophies.[71] From the first, Simpson saw the innovation of ether as one which would be driven by patients. Writing to his friend, Francis Ramsbotham, a London obstetrician who ardently opposed ether, in May 1847, he noted: 'I am etherising all my obstetric cases, the ladies all demand it here. Nothing but good results here.' A few months later, another letter told of the reluctance of London obstetricians to use chloroform which appeared 'To us in the North', as 'most frightful inhumanity. Here all our ladies demand relief – and quite right. Two or three medical recusants are found, but not many; and their patients desert them.'[72] Simpson understood the primary qualities of a practitioner to rest on a sympathy of suffering and so, he urged his students:

> the objects and powers of your art are alike great and elevated. Your aim is as far as possible to alleviate human suffering and lengthen out human existence. Your ambition is to gladden as well as to prolong the course of human life by warding off disease as the greatest of mortal evils; and restoring health, and even at times reason itself, as the greatest of moral blessings... If you follow these, the noble objects of your profession, in a proper spirit of love and kindness to your race, the pure light of benevolence will shed around the path of your toils and labours the brightness and beauty that will ever cheer you onwards and keep your steps from being weary in well-doing;... while if you practise the art that you profess with a cold-hearted view to its results, merely as a matter of lucre and trade, your course will be as dark and miserable as that low and grovelling love that dictates it.[73]

He was reading the fundamental quality of a medical practitioner in classical Enlightenment terms. Compare his view to those expressed by Edinburgh physician, John Gregory, in his *Lectures on the Duties and Qualifications of a Physician* which was published in 1772:

> I come now to mention the moral qualities peculiarly required in the character of a physician. The chief of these is humanity; that sensibility

> of heart which makes us feel for the distresses of our fellow-creatures and which of consequence incites in us the most powerful manner to relieve them. Sympathy produces an anxious attention to a thousand little circumstances that may tend to relieve a patient; an attention which money can never purchase; hence the inexpressible comfort of having a friend for a physician. Sympathy naturally engages the affection and confidence of a patient, which in many cases, is of the utmost consequence to his recovery. If the physician possesses gentleness of manners, and a compassionate heart, and what Shakespeare so emphatically calls the 'milk of human kindness', the patient feels his approach like that of a guardian angel administering to his relief.[74]

McCullough has carefully analysed how Gregory's philosophical premise for medical practice was informed by the concept of sympathy that was central to the writings of the Scottish Enlightenment philosopher, David Hume. In Hume's view, there was no hierarchical structure in the relationship between patient and practitioner; they shared and experienced the same emotions.[75] There are strong connections between this understanding and Simpson's approach to anaesthesia.

For Simpson and other Scottish doctors, anaesthesia was a vivid articulation of the humanitarianism which lay at the very heart of medical practice. It carried the same meanings for both patient and doctor: the relief of physical pain and suffering. Simpson viewed himself as a benefactor, providing therapeutic benefits (comparable to giving opium and other analgesia) and his conscious simplification of chloroform administration reflected a set of values which weighted heavily the 'bedside' qualities of the physician. To have portrayed anaesthesia as a complex or unpredictable procedure may well have been off-putting to patients and would have negated the dominant values. Simpson often explained that patients were frightened of inhalers but not of handkerchiefs, and we shall see in later chapters how patients continued to be fearful of technology throughout the nineteenth century. His method could also be read as one which was innately in tune with the individual patient. During the initial inhalation period, patients were often seen to turn their heads from side to side in an attempt to relieve the sensation of suffocation. Using a handkerchief enabled Simpson to simply follow their movements until they became unconscious – Snow frequently had to hold patients down during these stages.

Despite all the polarities in their practice of anaesthesia, there is no doubt that Snow and Simpson were each highly effective administrators and enjoyed a confidence and authority that few of their peers shared. Simpson's confidence arose from his belief that administering anaesthesia fulfilled the primary responsibility of a doctor: the practice of humanitarianism. It was a view which stressed the values of biographical medicine: the individual response of each patient to medical intervention. Snow's clinical efficacy can be understood to derive directly from science. He believed that practising within the safety parameters established by experimental work meant that patients remained safe from the risks of anaesthesia. Knowing that excited patients were only exhibiting the signs of the second degree of anaesthesia gave him the confidence to proceed steadily with an inhalation until complete insensibility was reached. His belief that sudden death under chloroform was caused by overdosage, which he had established through experiments on animals, was so strong that when, during the 1850s, he did suffer one chloroform fatality, he refused to accept the death was related to the chloroform; his certainty that he had not administered too large a dose enabled him to interpret the death as arising from the patient's fatty degeneration of the heart.[76] Snow argued that the different value systems of humanitarianism and science were not mutually exclusive. Knowledge of the effects of ether and chloroform freed doctors to act humanely, yet without compromising patient safety. Like Snow, Simpson suffered a chloroform fatality although he failed to publicise it, and there is some evidence that he became concerned about the safety of his method in the 1860s when he moved to using a specific face-mask, rather than a cloth or handkerchief.[77] Nevertheless, the Scottish method of giving chloroform remained popular with many doctors throughout the nineteenth century.

Snow and Simpson had died by the time the jubilees of ether and chloroform were celebrated in 1896/97 but commentators accorded both men the status of figureheads: Simpson's discovery and promotion of chloroform symbolised the humanitarian face of medicine; Snow's establishment of the principles of anaesthesia represented its scientific benefits. Nevertheless, Simpson's discovery of chloroform and his promotion of anaesthesia under the banner of humanitarianism won him far more public recognition and acclaim than did Snow's scientific practice. It suggests that patients continued to weight the values of biographical medicine and perhaps explains why, even in London practice under the banner of Snow, the use of technology was rare. We will see in Chapter 6 how Snow's successors drew on the bedside skills of reassurance and empathy to manufacture the specialist practice of anaesthesia.

# 4
# Risks of Life and Birth

By the early 1850s, medical and public attention was focused on the risks of anaesthesia. Patients' dread of surgery had metamorphosed into a particular fear of the anaesthetic process, especially the dangers of chloroform, and was heightened by a cluster of fatalities at the prestigious teaching hospitals of London in 1853–54. At the same time, the criminal use of chloroform in several widely publicised robberies and abductions created a public perception that the agent was potent enough to threaten social values of liberty and freedom; then it drove changes in the law. Chloroform emerged as a new social fear: it was a stark reversal of the liberating qualities to be found in the earlier frolics of ether and nitrous oxide.

In earlier chapters we have examined the broad issues of the introduction of ether and chloroform and the way in which doctors variously accommodated the process of anaesthesia within the new scientific medicine and the traditions of biographical medicine. Here, we focus on the question of risk. Several historians have previously addressed the debates on anaesthetic risk: Pernick argued that the utilitarian calculus of risks and benefits marked a new approach to medical practice in America;[1] Burney suggested that risk assessment emerged as the best means of 'managing the passions and anxieties provoked by anaesthesia';[2] and the controversies surrounding its use in childbirth have been analysed by Caton, Poovey and Youngson.[3] But there has been little engagement with the manner in which concern about the risks of the new process intensified other British debates of the 1850s.[4] What, for example, was the purpose of pain in surgery and childbirth? How should the public be protected from the dangers of poisons? Was birth a 'natural' event that doctors should leave well alone? This chapter traces the percolation of ideas of risk in anaesthesia through medical practice into the social domain where it became a lasting symbol of the darker side of progress.

Many issues emerge. Doctors sought to assess anaesthetic risk within established surgical and therapeutic frameworks, but their anticipation that anaesthesia would reduce surgical risk by removing the patient's fear of pain failed to materialise. Instead, a new risk emerged – patients' fear of unconsciousness – which was heightened by the associations drawn between chloroform and crime. In debates on childbirth, ether sustained arguments about female sexuality and impropriety, and chloroform fed into matters surrounding the status of birth. We shall see how, for all the rhetoric of risk, those labouring women who could afford its cost welcomed pain-relief as keenly as surgical patients.

## Quantification of risk

That anaesthesia provoked some of the most intense and controversial medical debates of the mid-nineteenth century cannot be doubted. Pernick's study portrays the risk assessment which accompanied its introduction in America as the means through which a new conservative medical ideology reconciled the opposing claims of the 'heroic' medicine of Rush and his followers, with the non-interventionist approach of natural healers. It arose, he suggests, because of the concurrence of anaesthesia with the rise of quantification of therapeutic safety and efficiency developed in Paris by Pierre Louis during the 1830s; the new numerical approach became the best way of resolving ethical and therapeutic conflicts.[5] But the British context was different. Since the early eighteenth century, quantification had been used to assess the value of surgical procedures and therapies, and Trohler's study shows that the 'principal actors were medical graduates of Edinburgh University'.[6] One of the first innovations to be subjected to such an assessment was smallpox inoculation. Trohler notes how numerical analyses became the basis of a 'heated controversy' which lasted for the best part of the eighteenth century.[7] The subsequent replacement of variolation by Edward Jenner's technique of vaccination in the 1790s produced a debate around the idea of risk – for example, the moral and religious risks of inoculating humans with matter from animals.[8] Thus in 1840s Britain, evaluating the risks of medical intervention against the benefits of alleviating suffering and preserving life was not a new phenomenon.

In 1845 Simpson had, unknowingly, foreshadowed the anaesthetic risk debates when he mused on the 'difficult moral and professional problem' of deciding whether or not to operate in cases such as ovariotomy, where the procedure was of immediate danger to the life of the patient. 'Am I', he said:

> conscientiously ENTITLED to inflict deliberately upon my own fellow-creature, with my own hands, the imminent and immediate chance of DEATH, for the problematical and prospective chance of his future improved HEALTH and prolonged LIFE?[9]

He warned that it was easy for 'the ideal glory of a successful operative result' to cloud the judgement of both patient and surgeon and so:

> with the patient the stern reality of danger and death too frequently vanishes . . . before the strong hope of life. And the surgeon. . . . in the computation of his successes is perhaps too liable to forget the actual amount of human suffering and human fatality through which these successes are obtained.[10]

The risks of anaesthesia to life were acknowledged in early debates on the use of ether, plausibly sustained by its chemical classification as a poison and the likeness of its effects to asphyxia. But as we saw in Chapter 2, concern focused largely on ether's violation of moral proprieties, rather than its physical risks. From the beginning, Snow and Simpson took the view that quantifying the risks and benefits of anaesthesia was simple. Painless surgery was not just of humanitarian benefit: removing the physiological risk of pain had, said Snow, 'the still greater advantage of saving many lives'.[11] Simpson attempted to quell opposition by asking: 'Does anaesthesia increase or decrease the mortality attendant upon surgical operations?'[12] He calculated that in British hospitals, pre-anaesthesia mortality rates for amputations of the thigh, leg and arm came to 29 deaths in every 100 cases. This was comparable to the range of mortality rates, 12–30 per cent, for major amputations published in 1844 by Thomas Inman.[13] The figures Simpson collected after the arrival of ether suggested that the mortality rate in British hospitals had dropped to 23 deaths in every 100 cases.[14] Six lives out of every hundred cases were being saved by anaesthesia, reasoned Simpson. Surely there could be no resistance to such utilitarian proof?

But although quantification was not a 'new' strategy in British practice, most doctors found that determining practice from statistics was intensely difficult. Few were prepared to privilege abstract data above their own experiences, despite the rhetoric of medical society debate which construed quantification as a progressive and laudable tactic.[15] Edward

Crisp, a physician to the Metropolitan Dispensary, was one of many who had taken an initial prejudice against ether; he had seen some animals die under its effects and had assisted with an operation during which the patient had an attack of asthma and died a few days later: 'never [having] recovered from the effects of the ether'.[16] Snow recognised that for doctors like Crisp the phenomena of anaesthesia were highly unnerving. He had sought to reassure readers of *On Ether* that anaesthesia was merely a chemical process:

> An appearance is met that would be truly alarming, if we did not know that it was only due to an agent which is flying away every moment in the breath, to leave the patient, in a few minutes, without any permanent trace of its having been there.[17]

But Crisp and his peers did not share Snow's confidence in the chemical principles of anaesthesia, nor did they have faith in statistics. Seeing or hearing of the wilting of life through the breathing of gas was reason enough to abandon its use; experiential knowledge remained the imperative of their clinical practice.

The switch to chloroform sustained a marked increase in the use of surgical anaesthesia: by January 1848 it was claimed that William Fergusson, London's leading surgeon, 'constantly' employed the agent.[18] Nevertheless, a persistent trickle of fatalities kept the risks of anaesthesia to the fore of medical and public attention. During this period, many doctors maintained that the dangers of anaesthesia were not new, instead they mirrored the inherent risks of therapeutic practice which derived from the 'peculiarities' of individual constitutions. Indeed, as we saw in Chapter 2, this was the explanation of Hannah Greener's death from chloroform in 1848. General principles could be established for the use of drugs, but it was impossible to predict the outcome in every case. The *London Medical Gazette* had summed up the broad principles for ether:

> in all persons the power of whose nervous and circulatory systems is distinctly impaired by age, in the subjects of manifest organic disease of the heart or aorta and in individuals who have an evident tendency to apoplexy, the use of etherisation is absolutely unjustifiable on the part of the surgeon, and eminently fraught with danger to the impaired organs and shattered constitution of the patient.[19]

Some like Robert Druitt, examiner at the Royal College of Surgeons, extended the exclusion zone to children and many were concerned that conditions such as epilepsy also increased the risk of fatalities.[20] Bristol surgeon, Augustus Prichard, who suffered a chloroform fatality during 'a very painful and possibly tedious operation' for excision of the elbow, sought to explain the cause of death in poison similes. The post-mortem had provided no clues, so Prichard reasoned that:

> to some constitutions chloroform acts as a poison, and that at present we have no means of knowing what are their peculiarities. The drug acts upon the heart in these cases...and we should therefore avoid its use when there are symptoms of diseased heart or brain.[21]

That therapies carried risks was well known amongst patients and doctors. Mercury had been used to treat syphilis from the sixteenth century and the growing use of opiates from the 1790s onwards had led to deaths from overdosage or bad effects which had been attributed to the particular susceptibility of the individual's constitution.[22] A strong analogy was drawn between anaesthetics and poisons – substances such as aconite, arsenic and prussic acid which could be of great therapeutic benefit if used in miniscule quantities, though in larger doses could prove fatal.[23] It was not possible to make chloroform an exception to the 'general principle' of therapeutic risk, said Edward Murphy, professor of midwifery at University College, in 1848:

> [chloroform] must disagree with certain constitutions . . . but it would be just as absurd to decide against chloroform on the strength of such cases, as it would be to prohibit opium, mercury or castor oil because they sometimes act violently on peculiar habits.[24]

But in the wider reaches of society a 'moral panic about the unrestricted availability of poisons' was developing.[25]

Public fear of poisoning had been growing since the early nineteenth century. Several well-publicised murder trials and the wide availability of poisons like arsenic and strychnine had created a perception that poisoning was the 'Crime of the Age'. The restriction of the open sale of dangerous drugs was part of a raft of legal and medical measures which sought to tackle the risks of poisons.[26] In 1851 during the peak of publicity about chloroform crime, a bill was passed to restrict sales of arsenic.[27] The measures won broad support from the medical community, and the *London Medical Gazette* commented that as well as preventing the 'numerous deaths' which occurred annually through either criminal intent or accident, the bill was a useful test-case for controlling other poisons such as chloroform.[28] And it was against this backdrop of concern about poisons that debates on anaesthetic risk were brought to a new intensity by a series of surgical chloroform fatalities in London in 1853–54.

## Death in London

The risk of death from chloroform was established from the beginning, and between 1847 and 1852 Snow had recorded a worldwide total of 29 fatalities – five of which had occurred in London practice.[29] However, between March 1853 and December 1854, there were a further seven chloroform fatalities in London practice alone – Appendix Tables A.9, A.10. That six out of the seven deaths had occurred in the prestigious teaching hospitals and in the hands of elite surgeons such as Thomas Bryant at Guy's and James Paget at St Bartholomew's was deeply disturbing. There was no discernible pattern to such deaths. In some cases chloroform was given in an inhaler, in others on a cloth. Some patients were nervous, others were not, and procedures ranged from amputation to the removal of a toenail. The fatalities challenged the assumption that the metropolis was the pivotal centre of progress – we shall see in the next chapter how London surgeons curtailed their use of chloroform during this period.

Widespread coverage of the fatalities appeared in general press and medical journals, and in June 1853 at a meeting of the Medical Society of London, Edward Crisp, a physician to the Metropolitan Dispensary, proposed that the 'general utility' of anaesthesia required 'serious consideration' by the profession. Its benefits were undeniable, he said, but the key issue was 'the preponderance of the good over the evil', in other words a balancing of risks and benefits. Crisp had published details of around 40 deaths that he believed had been caused by chloroform and drew on this evidence to recommend that members should exercise far more caution in their use of anaesthesia.[30]

Crisp's concern was echoed some months later by the *Lancet*. The chloroform fatalities had proved that therapeutic parameters were no longer sufficient to curb the risks of anaesthesia, said the journal; if aconite or opium had caused a similar number of deaths, then those drugs would have been withdrawn from the pharmacopoeias.[31] The *Lancet* recommended restriction because 'indiscriminate administration' had led to 'very grave abuses' and unnecessary deaths:

> had it been otherwise, had chloroform never been inhaled, save when its use was necessary, lives would not have been sacrificed to

> the removal of a tooth, a toe-nail, or a little finger, in tapping a hydrocele, or touching a sore with caustic.

Surgical anaesthesia could only be justified in those cases where either the 'intensity or duration of the pain' constituted a serious threat to life, or where the success of the surgical procedure was dependent upon patient compliancy. In all other procedures its use should be excluded.[32] Only weeks later, after a further chloroform fatality at Guy's, the *Lancet* railed again on the subject, this time asking:

> was the intensity or duration of the pain in an amputation of the leg sufficient to justify the fatal risk in such a subject? Or can it be said that insensibility was essential to the surgeon's proceedings? Surely not.[33]

This may well have been an extreme response to a situation that appeared to be rapidly spiralling out of control. But the suggestion that risk would diminish through restriction failed to take account of many of the 'givens' of surgery. In particular, the realities of pain and individuality.

## Individual sensibilities

Through the growing use of opiates during the 1830s and the 1840s, patients had become accustomed to a medical culture in which, in respect of chronic disease and death, pain-relief was understood to be a primary function of medical care. From a patient's perspective, the innovation of ether and chloroform appeared to extend these benefits to surgery. But for surgeons, the risks of anaesthesia created new ethical dilemmas. Did a surgeon have the right to deny patients the use of anaesthesia if he believed the risks of insensibility were greater than that of the surgery itself? If the patient insisted on using anaesthesia, was he (the surgeon) reneging on his professional duties by operating under such circumstances? Evidence suggests that many patients had determined upon the use of anaesthesia prior to any discussions with their doctor. One of the first London fatalities at St Thomas' hospital in October 1849 was a male porter who had decided 'before entering hospital' that he would have his toenail removed under chloroform.[34] In Edinburgh, the surgeon, James Gillespie tried to dissuade a 17-year-old girl from using chloroform for a tooth extraction. She died during the inhalation, and it later emerged that the reason she had gone to

Gillespie was because her previous dentist had given up using chloroform.[35] A further complication was the understanding that it was impossible to quantify individual susceptibility to pain.

When Sir Robert Peel, one of Queen Victoria's longest-serving politicians, had a riding accident in June 1850, he broke his leg so badly it eventually proved fatal.[36] Sir Robert suffered from 'excessive sensitiveness to pain' and this prevented his injury being examined, let alone treated; he died within a few days. Sir Benjamin Brodie, late President of the Royal College of Surgeons, and Caesar Hawkins, surgeon at St Georges' Hospital, attended Peel. Both surgeons had experience of using chloroform yet did not employ it on Peel.[37] Brodie later recalled that 'if he (Peel) had been brought under anaesthesia it might have been easy to have put the broken bone in place, or removed it, and if necessary secured the wounded vein'. But 'all this was out of the question at the time'.[38] We can assume from this that Peel was not offered chloroform. But why not? His accident occurred only days after a police constable died under chloroform at Guy's whilst having part of his hand removed. In that instance, the surgeon, Edward Cock, had tried to 'dissuade' the patient from having anaesthesia, despite the severity of the operation.[39] In the light of this event it may well be that Brodie and Hawkins were unwilling to take the risk of killing the ex-Prime Minister. It may also be that they continued to claim a physiological function for pain. Although Snow argued that it was in fact the weak or particularly sensitive patient that anaesthesia was ideally suited to protect, many doctors believed that blotting out sensations through chloroform deprived the body of its natural safety mechanism.[40] The tragedy is also revealing of the respect accorded to individual sensibilities.[41] Brodie and Hawkins did not override Peel's pain threshold by forcibly treating his injuries which suggests that individual sensibility to pain was taken as sufficient reason for non-intervention. It also supports the earlier argument that the use of anaesthesia at this time was only acceptable with the patient's consent and cooperation. Thus restricting anaesthesia or using a universal calculus of risk proved unworkable and most surgeons resorted to assessing the individual risk of pain-relief within a set of general principles. It was a strategy which was far more reflective of the philosophies of therapeutics than of Snow's 'new' science.

## The 'peculiarities' of bodies

The only aspect that was 'impossible' for a surgeon to calculate prior to an operation, wrote Henry Smith in 1847, was the manner in which the individual constitution would respond to the knife.[42] Smith had trained in surgery with Benjamin Travers at King's College Hospital, and in his discussion of the causes of fatalities in surgical operations he explained how this individuality could affect outcome: 'the effects of a severe operation will differ in different individuals in proportion to the state of health at the time, and the mental and physical endowments of their system'.[43] The mental state of the patient had long been acknowledged by doctors to be a crucial factor in determining the outcome of sickness and disease. In 1803, Manchester doctor, Thomas Percival had advised that:

> The *feelings* and *emotions* of the patients, under critical circumstances, require to be known and to be attended to, no less than the symptoms of their diseases... Even the *prejudices* of the sick are not to be condemned or opposed with harshness... they will operate secretly and forcibly on the mind, creating fear, anxiety and watchfulness.[44]

In surgery, the mental state of the patient was one of the areas to which the surgeon paid particular attention in his pre-operative assessment; fear was delineated as an explicit physiological force that could weaken the heart and also adversely affect the nervous system. Medical journals and surgical textbooks abounded with examples of patients whose fear or apprehension of surgery was connected to death for no clear pathological reason.[45] Smith warned that in severe cases, the patient's physical powers became so depressed that recovery was impossible and 'he dies the victim of his own mental uneasiness'. He welcomed the innovation of ether because the prospect of painless surgery removed a large part of this unpredictable risk – the patient's anticipation of physical suffering. The risk that remained derived solely from the patient's anxiety as to the outcome of a dangerous operation.[46] But Smith's explanation failed to note that the very process of inhalation, and of unconsciousness itself, had become a focus of dread for patients; they feared they might never wake again.[47]

## Risks in the mind

During the early months of ether, expressions of anxiety on the part of the patient were frequently linked to the practical novelties of the process. Ether was pungent and irritant, it caused patients to cough and choke, and complaints of feelings of suffocation were common. Many doctors took the view that the best way of abating apprehension was to educate the patient in the techniques of the new process, either through watching another individual inhale, or by the patient trying it themselves in advance of the operation. Indeed we saw in Chapter 2 how some doctors inhaled ether themselves so as to avoid charges of patient experimentation. The dentist, Robinson, who was about to give ether to a 'nervous youth' in 1847 during a tooth extraction, first gave it to his manservant to show his patient exactly what inhalation meant.[48] At St Bartholomew's, the surgeon Samuel Tracy realised that explaining the principles of ether in advance to a patient facilitated the administration. He therefore:

> allowed the patient to practise the inhalation a few hours before being placed on the table; thus testing its effects on the particular case, and at the same time removing the fear induced by its novelty at the time of the operation.[49]

These problems diminished with chloroform which was a far easier gas to breathe. But patient anxiety remained.

In 1848, Edward Murphy, professor of midwifery at University College, suggested that the notion of unconsciousness 'excites' in the minds of patients: 'a thousand apprehensions, lest anything unforeseen should suddenly happen to them while they are in that state, and, however groundless these impressions, they are difficult to overcome'.[50] Worries about the loss of self-control at the deepest level revealed the deep tensions in understandings of the symbiosis between mind and body. The mind controlled the physicalities of the body: it was the guarantor of individual morality and adherence to social norms. But within the surgical context, the long-established link between undue apprehension of surgery and unexplained surgical death had revealed the capacity of the mind to expose the body to additional risk. Anaesthesia appeared to be the ideal solution: by removing pain it would allay patients' fear of surgery. But in many instances the promise of insensibility proved to be incapable of barricading the body from the consequences of fear and dread – now the fear and dread of chloroform. Hence the *Lancet*'s warning that: 'if he [the patient] should feel any apprehension or gloomy forebodings, chloroform should be steadfastly refused'.[51] Reports of chloroform fatalities incorporated not only justifications for its use, but also detailed the patient's mental state: Hannah Greener was known to have had a 'dread' of the approaching operation and was sobbing at the start of the fatal inhalation; a lady who died in St George's in 1854 whilst beginning to inhale chloroform, was noted to be 'much agitated' and 'seemed to breathe little'; and Snow recounted a patient who had died on the fourth day after lithotomy, as having 'a strong conviction that he should not recover'.[52] In France, similar arguments were used to explain the death under chloroform of Madame Simon who was in a 'state of extreme agitation and expressed unpleasant forebodings' before her inhalation.[53]

The common reading of such cases implicated fear as the agent which heightened the dangers of chloroform. Sibson, the Nottingham surgeon, who, like Snow, understood chloroform to cause respiratory or cardiac death, suggested that chloroform was particularly dangerous in 'those cases where palpitation and dyspnoea are easily excited, either from abdominal distension or from mental emotion'.[54] Snow agreed that fear was indeed a physiological threat but explained his view that fear and the effects of chloroform were mutually exclusive; fear, which was a product of consciousness, could not be sustained once the patient had become unconscious:

> It has been said that chloroform ought not to be administered if the patient is very much afraid, on the supposition that fear makes the chloroform dangerous. This is, however, a mistake; the danger, if any, lies in the fear itself... Fear and chloroform are each of them capable of causing death, just as infancy and old age both predispose to bronchitis, but it seems impossible that fear should combine with the effects of chloroform to cause danger... Fear is an affection of the mind, and can no longer exist when the patient is unconscious... When chloroform has been absorbed in sufficient quantity to cause unconsciousness, fear subsides, and with the fear its effects on the circulation.[55]

It was because of the control of 'the will' over the patient's breathing, that Snow stressed that inhalation should not commence until the patient's apprehension had eased; success depended on the initial breathing being regular and tranquil, and if the patient was agitated, this could have dire consequences.[56] He recounted a case in 1855 of a 21-year-old man who was about to take chloroform before a tooth extraction and when Snow took his pulse:

> found it to be small, weak and intermitting and it became more feeble as I was feeling it. I told the patient that he would feel no pain, and that he had nothing whatever to apprehend. His pulse immediately improved. He inhaled the chloroform, had his teeth extracted, woke

> up, and recovered without any feeling of depression. Now if the inhalation had been commenced in this case, without enquiry or explanation, the syncope which seemed approaching, would probably have taken place, and it would have had the appearance of being caused by the chloroform, although not so in reality.[57]

Thus in the conscious state there was a direct link between the emotions and the physiological dangers; but the removal of consciousness broke this connection and freed the nervous system of its influences, allowing it to respond predictably to the chemical effects of chloroform. It was necessary to handle patients on an individual basis at the beginning of an inhalation – Snow found that well-brought up young ladies could be soothed by kind words whereas rough men needed to be held down. But once under the influence of chloroform, their bodies and minds became subject to the same analytical framework.[58] Indeed, Snow had noted early on that under ether it was possible to analyse the phenomena of the mind – sensibility and perception, for example; ether 'decomposed' the workings of the mind, he said, in much the same way as galvanism broke down chemical compounds for analysis.[59]

Snow recognised how problematic the phenomena of anaesthesia were, outside the context of the new anatomical and physiological frameworks. Fear and chloroform both caused death by syncope and so clinical observation alone could give few clues to the particular cause. Instead, the chain of events needed to be placed within the framework of established principles. In 1854 a death occurred in St George's hospital whilst Henry Potter, whom Snow had trained in anaesthesia, was giving chloroform with an inhaler. Snow participated in the post-mortem enquiries, testing blood samples and portions of the lungs and liver for the presence of chloroform. None was found and he noted that 'everyone in the operating theatre was a witness that the expiratory valve of the face-piece was not, at any time, more than one-third closed'. The conclusion that the patient had died from cardiac syncope that had been caused by 'mental emotion' was based upon these quantifiable facts. He continued: 'it is only the absolute knowledge, that any small quantity of vapour which this patient inhaled was very largely diluted with air, that enables one to decide with confidence, that the chloroform was not the cause of death'.[60]

In Snow's view, the suspension of the emotions and their power over the body through anaesthesia reduced each patient to a physiological organism. But another more sinister view was that emotions could have subversive influences on the experience of painless surgery, far beyond the obvious dangers of overdosage. Charles Dickens, writing in 1853 as editor of *Household Words*, pondered on the 'curious fact' that 'fatal results have often followed the administration of Chloroform to the persons who have exhibited decisive and unaccountable dread of it'. The only way in which this phenomenon could be understood, suggested Dickens, was as 'some theory or instinct or by some superstition of the forecast shadow of approaching fate'.[61] It was, perhaps, a fear which had been given shape and form through the emergence of chloroform as a new tool of criminal power.

## The criminality of chloroform

The potency of the new anaesthetic agents – ether and chloroform – not just to create insensibility, but also to kill, was well established in the public domain. Not only was there widespread coverage of fatalities in the press but accidents had occurred during popular demonstrations on the new technique. In 1848 at the Royal Institution, the chemist, Professor William Brande demonstrated the effects of chloroform to a very large audience which included several ladies. Unfortunately, the guinea pig which Brande used to show the manner of chloroform insensibility died in the vapour. It was hardly a positive image: every member of the audience 'would remember the fate of that animal, and dread its application to themselves', claimed the *Lancet*.[62] Snow too had experienced a similar event whilst demonstrating the effects of ether to medical officers from the military. Distracted by his lecture notes, he forgot to remove a thrush from the jar of ether in time to save it. He had the sangfroid to acknowledge the death and presented it as an example of the way in which the powers of anaesthesia should only be given to orthodox doctors.

During 1850, public attention was drawn to the darker powers of chloroform by a series of convictions in which criminals were alleged to have used the chemical to overpower their victim.[63] The use of drugs for such purposes was not new and since the 1820s there had been incidents of victims being stupefied by opiates as a preliminary to robbery.[64] But the disturbing aspect of the claims about chloroform was its potency: it seemed to overpower victims instantaneously and without warning (Figure 4.1). In January 1850, solicitor Frederick Jewett was walking along the busy Whitechapel Road when he felt someone touch his side and hold a rag in front of his mouth. He immediately became insensible and remained unconscious for around 10 hours. On waking, he found himself naked on a dirty bed, with a piece of rag thrown over him. His trousers were covered in mud and he had lost his watch, money and

DARING ROBBERY OF AN OLD GENTLEMAN NAMED "BULL,"
**By the Aid of Chloroform.**

*Figure 4.1* The new tool of criminals: aided by chloroform, thieves attempt to rob John Bull of unwelcome tax measures. *Punch* (1851). Reproduced by courtesy of the Director and University Librarian, The John Rylands University Library, The University of Manchester.

some of his clothes; he struggled to escape as the door of the apartment was locked from the outside. On the day of the hearing, Jewett was still 'delirious' and Margaret Higgins and Elizabeth Smith, two women 'of notorious character', were charged with robbery by means of chloroform. Higgins had received the chloroform from her partner, Gallagher, who had recently undergone an operation at the London hospital, where doctors gave him 'some stuff to send him to sleep'; he had managed to steal some of it before his discharge.[65] A month or so later, another woman of ill-repute, Charlotte Wilson, received a 10-year imprisonment for using some 'deleterious article such as chloroform' to overpower a gentleman prior to robbing him.[66] Such accounts fuelled the public image of chloroform's dangers and its links to crime.

On reading these reports, Snow wrote to the *London Medical Gazette*. His intention was to separate the facts surrounding the action of chloroform from those reported in the press. No one, he stressed, was capable of inhaling chloroform without an awareness of the process. 'The sensation of pungency in the nostrils and throat . . . is so strong and peculiar that no person can take a single inspiration without being aware that he is inhaling something very unusual', he continued, and suggested that anyone who was the subject of such an attack in a public street would 'instantly hold his breath, and use all his powers of resistance to repel the assault'. Most reports by individuals who claimed to have fallen in the street with no recollection of anything other than a handkerchief being applied to their face, he remarked, were likely to be 'the ingenious invention of the reporter', or possibly, they had suffered a fit. Rather dryly, he suggested that such a notion had gained 'general credence': individuals 'who have to account for being in disreputable places and company, instead of the usual excuse of having been dining out, will try to remember something of a hanky'.[67] He directly linked the inaccuracy of these reports to the public's false understanding of the chemical and physiological realities of chloroform.

A couple of months later, two further cases received wide publicity. The first, in April 1850, involved Charles Jopling who took his sweetheart down an alleyway and tried to dope her by covering her face with a hanky soaked in chloroform. She shouted for help and was rescued by a policeman. Jopling was remanded in custody and then released on bail, whereupon he neatly arranged events by marrying his victim. His new wife pleaded for his discharge, and after further remand and a severe lecture, Jopling was released by the magistrate, who judged that: 'if she [his wife] does not remain of the same forgiving disposition, [she will] doubtless find means of punishment as severe as any which the law had in store for him'.[68] The second case was far more serious and occurred one night in a hotel in Kendal when a robber entered a bedroom and tried to overpower the man with chloroform. After a noisy and violent struggle, assistance came. Again Snow responded, stressing the seriousness of this case: 'the thief in the dark, and without experience to guide him, could not have known when to stop in time to spare life if that were his intention'. For this reason, Snow understood this assault to have the makings of a murder, and concurred with other commentators that an 18-month prison sentence was insufficient for the severity of the crime.[69]

Snow's concern to limit the damage of such frightening incidents led him to establish a method of detecting the presence of chloroform in corpses. In May 1850 he demonstrated his apparatus at the Westminster Medical Society and noted that the London coroner, John Parrott, had sent him some body parts of a woman who had died in mysterious circumstances for testing.[70]

But fear of the criminal abuse of chloroform fed into the wider concern about the dangers posed to the public from poisons in general; and as publicity linking chloroform and criminality grew, the government took action.[71] Lord Campbell, who was appointed to the Court of Queen's Bench, drew up a bill for the prevention of offences, and cited chloroform in its fourth clause. It recommended that unlawful use of 'chloroform or other stupefying drugs' should be punishable with transportation or a minimum of 7 years imprisonment.[72] The Bill met with widespread approval within Parliament and the Act was passed in June 1851. For Snow, this course of action was a travesty of science and served only to inflate public fears about chloroform's powers.

During the passage of the Bill, Snow had written an open letter to Lord Campbell suggesting that the word 'chloroform' should be withdrawn. Prosecutions could still be made, he said; they would be covered by the general phrase, 'stupefying drugs'.[73] The *London Medical Gazette* dismissed his arguments as 'narrow', and cited the use of chloroform in an operation on the grizzly bears in London's Zoological Gardens as an example of chloroform's potency. Snow's retort was sharp indeed. He described how each bear had been 'secured by a collar, and held by two or three men, whilst the chloroform was given to it'; he asserted that they were entirely under control *before* being made unconscious.[74] His experience showed that chloroform could only be administered with consent or with force. Force alone was a criminal offence. Thus to specifically cite chloroform in such cases would only serve to deepen the current public apprehension. He was explicit in his letter to Lord Campbell that: 'to legislate on this matter would revive the groundless fears of the public which have subsided, or been allayed'.[75] But Campbell's dismissal of such arguments, although courteous, gave voice to the view that although a 'strong man' may well be able to resist submission by chloroform, weaker members of society needed protection from its powers. It echoed indeed Snow's arguments for the protective role of anaesthesia for weak and feeble patients.

Although Snow's arguments failed to change the Government's stance, his view that the use of legislation was an unnecessary infringement of individual liberties chimed with the later ones of the philosopher John Stuart Mill.[76] However, his primary concern that adverse publicity could diminish anaesthetic use was unfounded; by 1853 Armstrong Todd, lecturer on ophthalmic surgery at the Chatham Street School of Medicine in Manchester, noted that:

> even among the lower classes ... [the] administration [of chloroform] is almost invariably one of the great conditions on which a patient consents to submit to an operation.[77]

Nevertheless, associations were drawn between chloroform, fear and criminality. The most evocative example is found in the Sherlock Holmes story, *His Last Bow*, in which Lady Frances, who has been abducted, is discovered by Holmes and Watson in a coffin:

> we tore off the coffin-lid.... there came from the inside a stupefying and overpowering smell of chloroform. A body lay within, its head all wreathed with cotton-wool, which had been soaked in the narcotic.... with actual suffocation, and what with the poisonous fumes of the chloroform, the Lady Frances seemed to have passed the last point of recall. And then, at last, with artificial respiration, with injected ether, with every device that science could suggest, some flutter of life, some quiver of the eyelids, some dimming of a mirror, spoke of the slowly returning life.[78]

Doyle had added fear of chloroform to the persistent worry about premature burial; it was a potent cocktail, in which science smelled of crime. For the remainder of the nineteenth century and beyond, fear remained a spectre of the anaesthetic process.

## The risks of life

Laws against new dangers such as chloroform sprang from a public conviction that Victorian life had become far more risky.[79] Radical innovations of the nineteenth century, such as the steam engine, had revolutionised every aspect of life – work, travel, communications; but such progress had a darker side. The artefacts of progress brought new threats to life. William Farr drew attention to such risks in 1858:

> young ladies are dressed in elegant muslins, but muslins [are] still inflammable, which sometimes, alas! burn away their beauty and reduce them to ashes; lighthouses and lightboats are often unprovided, and are often of no avail. Men are destroyed by explosions in mines for the want of adequate ventilation; by defective machinery; and by carelessness. In factories death arises from unfenced machinery. Railway accidents are sometimes accidental, sometimes the consequence of bad management ... the public ... should be taught the nature and the extent of all the dangers by which they are surrounded; for some of those dangers they will learn to avoid, and many of them can be diminished or entirely removed.[80]

Farr claimed that many innovations had been taken up before proper parameters had been established to limit their danger. He was positive that the risks to life from progress could be managed, partly through legislative control and partly through increased individual awareness.[81] Snow understood the problems of anaesthesia to have arisen in a similar way: the speedy take-up of ether and chloroform had allowed no time for the establishment of boundaries of safe practice. Many doctors practised anaesthesia outside the anatomical and physiological frameworks that gave meaning to the process, and for this reason the risks to life remained. Like Farr, Snow was optimistic that as the principles of sound practice spread through the medical community, risks could be substantially diminished. In his own practice he was confident that his adherence to principles established through experimental work on animals could eliminate risk. This was not the case however for amylene.

## Death from amylene

Amylene – made by distilling amylic alcohol with chloride of zinc – had been discovered by the French chemist M. Balard in 1844 and Snow learnt of it in 1856. During preliminary trials on animals he found it had many positive features; the main one was that although amylene had the same potential as chloroform to cause fatalities through paralysis of the heart in high dosages, it was much harder to achieve this and therefore it appeared a 'safer' anaesthetic. Snow placed it midway between ether and chloroform; he never changed his view that ether was the safest anaesthetic. The benefits for patients were significant:

amylene was an easier gas to breathe, consciousness returned more quickly, and the greatest advantage was the lack of post-operative vomiting. But out of around 230 or so administrations, he suffered two fatalities.[82]

The first amylene death occurred on 7 April 1857, only days before he was due to attend Queen Victoria for the second time. He went to Regent Street with William Fergusson who was to operate on Mr Wellington from Liverpool – 33-years old and 'a healthy-looking, well made man'. He had recently returned from Australia and needed to have an anal fistula repaired; he had previously had an operation in 1851 but the complaint had returned. The operation proceeded and as Fergusson was washing his hands at its conclusion, Snow drew his attention to the condition of the patient:

> he was...livid and the breathing of a very gasping character. The breathing left off, except deep, distant, gasping inspirations, and we performed artificial respiration, first by rolling in the method recommended by Dr Marshall Hall, then by pressing on the chest, the face being turned to one side... [we] continued for an hour and a half without effect... there was no remaining sensibility.[83]

The death of Wellington was a tremendous blow. He was Snow's 144th amylene patient and there is no acknowledgement of any previous problem in Snow's published work, although his notes suggest there was cause for concern during two administrations in January 1857: a 35-year-old woman had an impalpable pulse for a few seconds, and a 11-year-old boy became pale and livid and his pulse was also lost for a short period.[84] The post-mortem on Wellington revealed emphysema in the lungs which was believed in some way to have obstructed the pulmonary circulation. Snow continued to use amylene, saying later that during the period between this death on 7 April and the second fatality which occurred on 30 July: 'in the ninety cases and upwards in which I administered amylene between these two accidents, I never had occasion to feel a moment's uneasiness about it'.[85] After the second fatality occurred, Snow concluded that amylene's ready volatility, which could produce a variation in its boiling point, had produced a vapour which was too concentrated and thus caused cardiac death.[86]

Certainly Snow was not alone in his use of amylene; it had become popular with several French and German doctors. But a mark of the respect accorded to his anaesthetic skills came from the French Académie de Médicine which, on hearing of Snow's amylene fatalities, recommended its disuse. Their position was clear; if Snow had experienced problems, other doctors should leave well alone.[87] Benjamin Ward Richardson, his friend and biographer, claimed that Snow had not realised the danger until he [Richardson] had 'ventured to show him separation of amylene in the blood, a separation which looked like the formation of minute plugs'.[88] Richardson also claimed that after the two deaths in 1857, Fergusson, who provided Snow with the majority of his work, questioned him closely about his trialling of amylene and told him he had made a serious error in not subjecting it to more animal experiments before extending it to clinical use. At the time Snow had records of over fifty experiments on cats, guinea pigs, mice and linnets and well over a hundred successful cases behind him. Was Fergusson perhaps being rather unfair? His comments may well have been prompted by a concern for the impact upon his own reputation. Nevertheless it is possible to detect Snow's frustration that despite his best calculus of the risks of amylene he had suffered deaths.

In his first publication on amylene in January 1857, Snow noted that although the gas could not be accorded the 'absolute safety' of ether, he trusted that 'it will be perfectly safe with careful management'. From his experimental work he had gathered evidence that 'the cold produced during its evaporation would, in all the ordinary methods of inhalation, prevent the air from taking up a quantity of the vapour which would be dangerous'.[89] It seems that his decision to employ amylene arose primarily from its clinical benefits; there was never any suggestion that the depth or quality of insensibility was superior to that produced by ether or chloroform. Even after the first fatality – the death could not be attributed 'to any other cause than the amylene', he said – he still believed its benefits outweighed its risks. After announcing his second fatality Snow was criticised by Augustin Prichard, president of the Bath and Bristol branch of the British Medical Association, for falsely stating amylene's safety. Snow dismissed Prichard's claims; Prichard had 'entirely mistaken' his earlier statement on amylene. Yet the accusation hurt: 'I doubt whether the style of sarcastic reprimand, if not exultation, which he [Prichard] has employed, would be calculated to encourage other laborious attempts to advance the science and practice of medicine.'[90] Although Snow wrote of making amylene 'absolutely safe' by putting a set volume of amylene in a bag with air, so as to reduce the danger of its variable volatility, he recorded no more administrations in his casebooks and from this point reverted to chloroform.[91] Richardson remarked that the amylene fatalities: 'affected him very seriously, and his sudden and early demise may, in some measure, be attributed to their effects upon him'.[92]

## Propriety and physiology

Childbirth presented a very different context to that of surgery. The pains of childbirth had long been rationalised as a Divine reminder of God's displeasure with Eve's disobedience in Eden: birth was viewed as a 'natural' physiological event into which doctors intervened only as a last resort.[93] Although the use of forceps had been introduced in the eighteenth century as an aid for difficult births, only 1–5 per cent deliveries during the 1850s required medical intervention.[94] The role of the doctor was to watch and wait: surgical patients could choose to refuse operations and tolerate their condition, but the process of birth was non-negotiable. Without reliable contraceptive means, women had no autonomy over their fertility and were thus bound to live through frequent cycles of pregnancy and birth.[95] Between 1851 and 1860, maternal mortality was just under 1 death in every 200 births and it was to increase over the century.[96] Such statistics lend weight to the view that the process of birth was understood by most Victorian women to pose a substantial risk to their life and subsequent health. The fear that confinement would end in death was common and mothers-to-be frequently arranged all their domestic affairs before the birth.[97] Many women also feared that the pain of labour would be too severe for them to bear and they would be forced to cry out in a way that laid them open to charges of impropriety and moral weakness. Loudon suggests that prior to ether, opiates were used to provide pain-relief in labour, although the practice was given little mention in obstetric texts.[98] As in surgery, the free use of opiates were considered to add to the risks of labour by promoting foetal asphyxia and haemorrhage, and reducing contractions. It is of little surprise then that many women were strongly receptive to anaesthesia. Certainly amongst the middle and upper classes, women like Emma Darwin, Catherine Dickens and Queen Victoria were eager to use the new pain-relief and were supported by their husbands and by advocates like Snow and Simpson. But for doctors like the London obstetrician W. Tyler Smith, the innovation of ether appeared to violate the fundamental social and moral values invested in women as wives and mothers.

The practice of obstetrics had been subject to controversy since the rise of the male midwife in the eighteenth century. Many of the medical elite argued that it did not warrant the status of medicine and surgery; they saw it as a 'messy and unscientific activity' which was practised by untrained midwives and general practitioners.[99] Their arguments focused on the status of birth as a 'natural' event which did not require intervention, and upon the moral dangers of introducing a male presence in the delivery room. Opponents of anaesthesia drew on these arguments and recast the dangers of sexuality in terms of ether, rather than the male presence itself. (There was no acknowledgement that they were indeed beneficiaries of the earlier debates.)[100]

Tyler Smith and other opponents saw the process of labour as heightening the sexuality of the women; using an analogy to animals, they claimed that 'an erotic condition of the ovaria is present during parturition, and that sexual congress and conception may take place immediately after delivery'. But in women, the physical pains of labour protected them from this 'level of the brute creation' by neutralising the overt sexuality of birth. Ether, however, jeopardised these inherent checks. It removed the veneer of self-control which distinguished labouring women from animals and rendered them open to exchanging 'the pangs of travail for the sensations of coitus'. Ether was anaesthetising not just physical pain, but the very values of 'chastity of feeling' and 'emotional self-control' that defined femininity.[101] Women depended upon doctors to help them preserve these objectives, argued Smith, and the employment of ether contravened fundamental values in the relationship between the labouring woman and her practitioner. He was not alone in configuring his opposition to the use of ether in terms of the wider issues of morality and sexuality.

George Gream, medical officer at Queen Charlotte's lying-in hospital, wrote of the ridiculous manner in which some doctors (those such as Simpson) had tried to 'persuade women that they have a right to insist' on the use of pain-relief. 'A stronger feeling was never evinced against anything medicinal', he warned, 'than is shown by the great body of English women against the innovation that has been attempted to be forced upon them'.[102]

For Simpson, however, it was precisely on account of women's heightened sensibilities that they suffered so intensely during childbirth. Women in less-cultured races had less-acute sensibilities and therefore bore the process of birth more easily, as did the less well-educated and refined women of Britain.[103] In 1842, the Manchester surgeon Charles Clay had spoken of his observation that labours amongst the 'factory-working population' were often accompanied with 'so little emotion or apparent suffering... as to border on unconsciousness, [or] at least extreme apathy'.[104]

Simpson justified his support for childbirth anaesthesia on humanitarian grounds; the pain of 'common labour', he said, was 'as great, if not greater, than that attendant upon most surgical operations'.[105]

Although many historical accounts suggest that a key element of the debate was religious opposition, this appears not to have been the case. Simpson did indeed publish a pamphlet on the religious justification of anaesthetic use in childbirth and surgery, but this seems to have been in anticipation of opposition, rather than in direct defence.[106] He also quoted statistics collected by Dublin accoucheur Joseph Collins from the Dublin Lying In Hospital, which suggested that maternal mortality was severely affected by the length of labour: 1 in 320 mothers died when labour was completed within 2 hours; 1 in 6 mothers died when the process exceeded 36 hours.[107] Ether then also offered physiological benefits.

Such benefits, said Protheroe Smith, lecturer in midwifery at St Bartholomew's, arose from ether's 'action on the various portions of the nervous system, and consequently, the effects produced by it on the function of parturition'. Smith claimed success in all his cases with not one 'unpleasant symptom'; ether was an aid in placenta praevia, turning, perforation and forceps cases: 'in short, some of the most formidable exigencies which the obstetric practitioner is called upon to meet'.[108] Smith claimed priority in the use of ether (and subsequently chloroform) in childbirth in England. He understood his use of the technique to be progressive, and of distinct advantage in attracting patients.

Nevertheless, even advocates of ether expressed disquiet at its effects in labour. Edward Murphy, professor of midwifery at University College and a friend of Snow, although highly positive about ether's power to facilitate difficult labours – those involving forceps or other techniques of intervention – struggled to come to terms with the transformation of the patient's state. He was dissatisfied 'with the manner in which patients had recovered from its effects; the state which they then exhibited was similar to half-drunkenness'.[109] This did not mesh well with Murphy's expectations of labouring women. After Simpson's announcement of chloroform, Murphy immediately substituted it for ether and found to his delight that 'the patient awoke quite tranquil'.[110]

Whereas Tyler Smith had found that ether placed labouring women outside propriety, Murphy described chloroform as its alter ego; its properties tamed women and brought them within social norms. In August 1848 Murphy attended a first time mother but as the severity of her pains increased 'she became very boisterous, so much so that sometimes her pains were cut short'. Murphy resolved the situation by offering chloroform:

> she willingly inhaled it, and felt so much relief that I could hardly get the inhaler from her hand. Not wishing, however that she should inhale more than I pleased, I did not renew the chloroform, but left the inhaler with her as long as she amused herself with the belief that it was doing her good. However, she was soon undeceived by the increasing intensity of her sufferings, and earnestly entreated for more chloroform; more was added but only on condition that I should have the inhaler in my own hands, to which she consented.[111]

On another occasion, during the labour of the wife of a medical practitioner, Murphy reported that at the point when: 'the pains were becoming perfectly intolerable – she was beginning to lose all command of herself; . . . for the first time I proposed to administer chloroform'.[112]

Chloroform was clearly effective in producing a state during labour which was at one with social aspirations. Snow described how relatives or friends who entered the labour room not knowing that the patient was under the influence of chloroform immediately 'praise the unconscious patient for her fortitude'.[113] But such medical control over childbirth could not be exercised without an equal expertise over the administration of chloroform.

During the first months of ether, practitioners like Simpson who employed it in childbirth induced full unconsciousness in the patient for the duration of the delivery. But Murphy and Edward Rigby, lecturer in midwifery and women's diseases at St Bartholomew's hospital, found that it was possible to provide enough pain-relief without removing the consciousness of the patient.[114] In Snow's classification of the degrees of anaesthesia, this was the second degree of anaesthesia; a condition in which 'the patient . . . has no longer a correct consciousness of where she is, and what is occurring around her'. It was a state which not only diminished the pain but also maintained medical control over labour:

> In this state, the patient will sometimes assist the labour by bearing down voluntarily, if requested to do so, and be otherwise obedient to what is said; and by withholding the chloroform for a few minutes, she at any time becomes quite conscious.[115]

To achieve a state that was painfree yet without full unconsciousness required a different approach to that of surgery. Snow noted that it was necessary for the chloroform to be administered 'very gently' in order to minimise any mental excitement and for the doctor to have no conversation with the patient. Whereas in surgery, such excitement was easily quelled by pressing on to full insensibility, when only the second degree of anaesthesia was required the process had to be more controlled.[116]

That chloroform pleased patients and doctors ensured its rapid acceptance in difficult labours. Doctors were also able to provide statistical evidence of its benefits: Murphy noted that in his practice, the use of anaesthesia in 59 'difficult' labours produced a mortality rate of 1 in 7 which was considerably better than the accepted average of 1 in 4 for labours using forceps.[117]

Opposition to pain-relief in childbirth did not cease on the switch to chloroform but opponents shifted from propriety to physiology. Tyler Smith suggested that the woman's response to the pains of labour was in fact of physiological benefit; the 'extra-uterine pressure' of the uterus was relieved by the opening of the patient's glottis which occurred during screaming or shouting.[118] And physician Samuel Merriman drew on old arguments that had accompanied the innovation of male birth attendants, to suggest that by intervening in 'normal' labours doctors were trespassing in the domain of nature:

> 'normal' labours which justified the use of chloroform were very few, and, scarcely come under the limits within which interference with nature can be allowed with impunity. The reasons, therefore, actuating the physician to allow the inhalation of chloroform in these simple cases must be exceedingly strong, or he will violate the law of non-interference with nature, founded on the experience of so many physicians of celebrity during a succession of years; and although he may not notice any immediate ill consequences, he must expect to find some sooner or later.[119]

In justification of 'normal labour', Tyler Smith and Merriman clung to the view that the pain of contractions was functional, but the new understandings of the nervous system suggested that sensation could be suspended without adversely affecting the reflex action of the uterus. The process of labour had two parts: the muscular contractions of the uterus and the sensations of pain, explained Isaac Baker Brown, surgeon-accoucheur at St Mary's hospital. Chloroform worked by removing the sensations of pain whilst leaving the uterine contractions unchanged. Nevertheless, Brown was reluctant to use it in 'common cases', a reflection of the way in which chloroform in practice appeared to diminish contractions which could either impede the birth or lead to the hazard of a retained placenta.[120] Snow claimed that although on a few occasions chloroform did appear to quicken or retard labour, it was not a statistically valid risk. He remained unconcerned by such possibilities. If chloroform did retard the labour by diminishing the strength of uterine contractions it was 'a matter of no consequence, however, as the patient is not suffering in any way'.[121] And he continued:

> the determination of the kind of labours in which chloroform should be used, or withheld, is really a matter of not much importance, because, as we pass from cases that are severe and protracted to those which are short and easy, the quantity of chloroform that is used, and the amount of diminution of the common sensibility, and of interference with the mental functions, become so trifling, that very little remains about which to hold a discussion.[122]

Nor too was he concerned about the possible placental transmission of anaesthetics from the mother to child, although on several occasions he had noted the smell of ether on the breath of newborn babies.[123]

As in surgical anaesthesia, many of the initial concerns about the risks of pain-relief in childbirth lessened as the process became familiar to both patients and doctors. By 1855, 'the use of chloroform in midwifery has become general', affirmed Murphy. That most doctors eventually adopted the use of pain-relief in childbirth, despite their reservations, is shown by the way in which its strongest opponents – Tyler Smith and Gream – used it in their practices.[124] They chose to accommodate the risks of anaesthesia in childbirth rather than risk losing patients.

Aside from attracting patients, many accoucheurs understood chloroform to provide a further opportunity for distinguishing their practice from that of midwives. Although the sale of chloroform was not subject to special conditions, like those introduced for arsenic in 1851, from the beginning doctors had argued that it could only be used safely under medical authority, a view reinforced by the public perception of chloroform's potency established in the criminal trials of the early 1850s. Although in practice the attending nurse or midwife often gave the chloroform, this was at the instruction of the attending doctor.[125] William Ackland, general practitioner in Bideford, Devon, stressed in his notebook that the quantity of chloroform to be added to the inhaler during its use in childbirth 'must be left to the judgement of the medical man'.[126]

Indeed the only death in Britain to be associated with the use of chloroform during childbirth occurred in 1855 when there was no doctor present. The mother had used chloroform in a previous labour in America but had been refused it by her English accoucheur who was also a family friend. Unbeknown to him, she obtained chloroform and it was given by the monthly nurse during her labour. The accoucheur (who was sleeping in the house at the time) was not called until the patient was dead. Snow observed that the death appeared to have taken place very slowly: 'the monthly nurse was extremely stupid to allow the patient to die'. Nor would the accident have occurred, he said, had it not been for the accoucheur's 'extreme objection' to chloroform.[127]

The use of chloroform for pain-relief in labour was successful on many different levels: it empowered the labouring woman to manifest her aspirations of self-control; it supported the role of the obstetrician, rather than that of the midwife; and it demonstrated vividly, the manner in which doctors could intervene into the natural processes of the body to ameliorate the experience of birth. Even so, whether or not a mother was given chloroform depended heavily on her finances.

## Labouring women

Snow had gained experience in childbirth through his care of patients in his general practice in the Soho area of London. After the innovation of ether in 1846, he had continued to run his general practice, as well as establishing himself as a specialist anaesthetist. His remaining casebooks, which run from 1848 to 1858, show that he attended around 157 deliveries in these 10 years, some in the capacity of accoucheur, and the remainder as the administrator of chloroform – Appendix Table A.11. Some historians have argued that care in childbirth was structured according to class, with the poor and working classes using female midwives, the middle classes opting for general practitioners, and the upper classes obtaining care from physicians. The trend within the middle and upper classes does seem consistent with the available historical evidence, but there is much to suggest that the distribution of care to the lower classes was far more complex.[128] Although we have no information on Snow's obstetrics practice prior to anaesthesia, we know that between 1848 and 1858 he attended 71 working-class mothers. His records suggest that they kept their antenatal care to a minimum, and in many cases he only attended the actual delivery. Postnatal contact was more frequent, but was primarily determined by the health of the mother and baby. Snow often attended the mother in subsequent labours, and developed a thorough knowledge of her obstetric history. Thus, although his working-class patients may well have sought to economise on costs by restricting their access to Snow to a minimum, they had chosen to use a doctor, rather than a midwife. We see from the following examples how his use of chloroform was patient-led.

On 28 December 1848, Snow attended Mrs Duhy. Labour had started and she was 'distressed' and suffering 'severe pains'. He dosed her with laudanum which relieved the pain and gave her a peaceful night. The pains returned in the morning and when he returned again in the afternoon, she was 'screaming out and complaining bitterly of her suffering'. At this point he gave her chloroform, which eased the pain, and the child was born some time later. Although the baby seemed well at birth, it died after only eight hours.[129] Less than a year later, Mrs Duhy was in labour again. The baby was in an arm presentation, and chloroform was used to keep her fully insensible whilst the baby was turned and then delivered in a breech position. Despite attempts at resuscitation, the baby died after half an hour.[130] Finally, Mrs Duhy was delivered of a healthy baby girl in February 1851.[131] Another patient, Mrs Terry of Frith Street, was 'weary and impatient' during a long and tedious labour. Chloroform provided relief until the child was delivered.[132]

But although Snow was willing for his use of chloroform to be patient-led, it seems likely that for poorer mothers this was a choice determined by economics rather than sensibilities. We have no figures on Snow's fees but most doctors charged patients additionally for the use of chloroform, as they did for other therapies. Estimates suggest that the use of chloroform may well have increased the cost of delivery by around 30–50 per cent, when normal midwifery charges ranged from fifteen shillings to five guineas.[133] That chloroform was used in only 11 per cent of the births of his general practice patients suggests that either these were labours where the mothers were able to bear the pain 'cheerfully', or it could suggest that the additional costs of pain-relief were too great for many working-class families to sustain. Snow did occasionally attend women in the local workhouse to give them chloroform, but on these occasions he was called in by the workhouse doctor, John French, and the births were difficult and required forceps. It would appear then that for many working-class women, the additional cost of chloroform put pain-relief beyond their budgets. For those who could afford it though, it became most desirable.

In the deliveries which Snow attended in his specialist capacity, most were straightforward, normal labours without complications, but he was called in by the attending accoucheur to administer chloroform. These patients, of course, were from the middle and upper classes.[134] The most obvious examples are Queen Victoria's two labours which he attended in 1853 and 1857; from an obstetric perspective these were normal, problem-free deliveries. Her use of chloroform is a good example of patient demand.

## Royal prerogatives

It was well known throughout the Queen's network of family and friends that she had an abhorrence of pregnancy and childbirth: following Simpson's announcement of chloroform, she was sent a copy of his pamphlet by her close friend Harriet, Duchess of Sutherland. When the Queen replied to thank her, she mentioned a mutual acquaintance, Lady Hardwicke, who had been one of the first ladies to use chloroform during her confinement that December.[135] From this point onwards the Queen was determined to profit from the benefits of chloroform, but it took 5 years to convince her medical entourage that it was safe. Gradually, through watching Snow administer anaesthesia, the Queen's advisers, physician Sir James Clark and accoucheur Charles Locock, developed a confidence in Snow's skills.[136]

The first report of Snow's attendance on the Queen, on 7 April 1853, appeared in the *Association Medical Journal* several days after the birth. The editor spoke of the Queen's excellent recovery and affirmed Snow's skill and experience in anaesthesia. It was, he proclaimed, 'an event of unquestionable medical importance', and the journal was very hopeful that it would remove the 'lingering professional and popular prejudice against the use of anaesthesia in midwifery'.[137] The *Medical Times and Gazette* too was supportive of chloroform and called for it to be given to all mothers in labour.[138] The *Lancet* however protested that

> in no case could it be justifiable to administer chloroform in perfectly ordinary labour; but the responsibility of advocating such a proceeding in the case of the Sovereign of these realms, would indeed, be tremendous.

Chloroform was of 'immense importance in surgical operations', it continued, but condemned the 'dangerous practice' of using it during 'a perfectly natural labour'.[139] That the *Lancet* was speaking for some practitioners, if only a minority, was reinforced by a letter from Dr Sheppard, a physician living in the provinces who wrote to the *Association Medical Journal*. 'No female for whom I have any regard shall ever, with my consent, inhale chloroform', vowed Sheppard: 'I look upon its exhibition as a pandering to the weakness of humanity, especially the weaker sex.'[140]

Snow did not enter into any of this public debate but sat down on 1 June and wrote a most authoritative paper on all aspects of the administration of chloroform during childbirth. 'I believe', he said, 'that no one disputes the power of chloroform to relieve the sufferings attendant on parturition . . . the benefits arising from chloroform in severe cases of labour are experienced in a lesser degree in favourable cases; and the patient may be fairly allowed to have a voice in this, as in other matters of detail'.[141] Four years later though, after Snow had attended the Queen during the birth of Princess Beatrice, it was clear there had been a significant shift in attitudes to childbirth anaesthesia and the *Lancet* reported in dulcet tones that:

> the labour was in every respect natural, as was the presentation . . . the pains were somewhat lingering and ineffective and so it was thought desirable that chloroform should be administered . . . the anaesthetic agent perfectly succeeded in the object desired.[142]

Thus by the late 1850s the risks of anaesthesia in surgery and in childbirth had, by and large, been accommodated within practice. The concern generated by the cluster of London fatalities in 1853–54 had diminished and the lack of fatalities associated with the use of chloroform in childbirth had established its use in normal labours. In its review of Snow's book *On Chloroform* which was published after his death in 1858, the *Lancet* recalled 'the senseless outcry that has been raised against the use of chloroform, on account of certain fatal accidents' and stated confidently that Snow's work showed how these were 'preventible' and had arisen 'from a want of caution in its use'. The risk of occasional chloroform fatalities was considerably less than many other 'powerful remedies', it said, and estimated that its use had saved the death of 1 in 100 patients undergoing surgery from 'shock, and the terror and pain of a severe operation'.[143] But for many doctors, although anaesthesia had reconfigured the surgical debate, its substance had not changed: the power of the emotions to endanger the body's physiology remained strong, even in an unfeeling body.

# 5
# Anaesthesia in London: John Snow's Casebooks

Some historians have claimed that aside from humanitarian benefits, anaesthesia did not affect surgery in any significant way until surgeons had gained control over the problem of infection through antisepsis techniques in the 1870s. Greene, for example, compared the types of operations performed between 1846 and the 1870s and concluded that anaesthesia had 'little immediate effect' on the development of surgery.[1] It is certainly true that the use of ether and chloroform did not remove the risk of post-operative wound infection, and surgical mortality remained relatively unchanged until the 1870s. But, as this chapter will show, by providing a solution to the problem of surgical pain, anaesthesia changed surgical thinking and practice, and revolutionised patient attitudes to operations.

Whereas previous chapters have concentrated on the intellectual arguments that accompanied the introduction of anaesthesia, here we focus on the fine detail of everyday London hospital and private practice. Although data on the early use of anaesthesia is hard to find, we are fortunate in having a particularly rich source of material in the form of three casebooks kept by Snow.[2] His records provide a 'cameo' of the first decade of anaesthetic practice. They cover almost 4500 anaesthetic administrations, with information on the anaesthetic and techniques used, the location of the operation, the name of the surgeon and the surgical procedure. He also recorded the response of patients to the innovation: those he had to coerce into using it, those who demanded it and a small number who refused it. The records begin in July 1848 when chloroform was established as the dominant anaesthetic, and they run to the time of Snow's death in June 1858. Also included is general practice data which shows very clearly how Snow metamorphosed into an anaesthetic specialist within the first couple of years of chloroform use. The casebooks yield a wealth of data on what Snow actually did; they represent the actualities of practice as he saw it, rather than the polished rhetoric of published articles and books. They provide a rare opportunity for historical analysis of practice against the narrative of publication. Snow worked for the majority of the elite surgeons and dentists of the metropolis. He gave anaesthetics to patients in many of the London teaching and specialist hospitals, to private patients who had operations performed in their homes, lodgings or hotels, and to numerous dental patients. He was acknowledged to be the most skilled administrator of the period and instilled confidence in surgeon and patient alike. Given his conviction that any patient fit for surgery was fit for anaesthesia, it is reasonable to assume that his practice illustrates the breadth and scope of anaesthetic use in surgery and dentistry in London during the 1850s. Where possible, the patterns revealed by Snow's data have been compared and contrasted with material gathered from the surgical records of London hospitals, particularly King's College, Guy's, the London, St George's and St Bartholomew's hospitals.

By 1846, each hospital kept records covering patient admissions, discharges and deaths.[3] These were usually split into surgical and medical patients, and then recorded under the names of individual wards or admitting surgeons. Some hospitals, such as the London, kept operation logbooks; others, like St Bartholomew's, recorded operation details in surgical casenotes. Operation notes had become broadly standardised by this time and contained a set of core data – age and sex of patient, history of illness, diagnosis and details of the surgical procedure. Therapeutic treatment was usually recorded in the ward casenotes, and if stimulants such as opium or alcohol were given during surgery or after an operation, then their use was included in the operation notes. But the recording of innovations such as anaesthesia was arbitrary – as was the noting of antisepsis routines in the 1870s. The recorded detail on the use of ether or chloroform varied enormously and depended very much upon the specific interest of the particular dresser compiling the notes.[4] By the 1860s the innovation was established enough for the use of anaesthesia to have become part of the core operative data, but it was not until the end of the nineteenth century that hospitals began to keep separate anaesthetic records. I have therefore been cautious in my extraction of data from hospital records and have taken the view that prior to the early 1860s, the employment of anaesthesia cannot be presumed unless there is specific reference to its use.[5]

Chapter 1 established the emergence of conservative surgery from the 1820s onwards and, in the absence of pain-relief, surgeons' expectations of patients' self-control. Here we begin with a sketch of London surgery c.1846 which reveals the landscape to which ether and chloroform were introduced. We then turn to the detail of the first 15 years or so of anaesthetic practice in London; a period during which the long-term viability of painless surgery and dentistry remained controversial, and the risks were often perceived to outweigh the benefits. Although perceptions of risk shaped practice, as we shall see through surgeons' response to the 1853–54 cluster of fatalities, by 1860 anaesthesia had transformed surgical practice. It promoted the practice of conservative surgery and reconstructed the criteria used to determine the propriety of operations. The number of operations performed in London practice had increased; more women and children appeared on hospital operating lists; and in private practice, middle-class women emerged as the new consumers of painless surgery and dentistry.

## Surgical practice c.1846

By the 1840s, London housed 10 key teaching hospitals, numerous dispensaries and a growing number of specialist institutions.[6] 'There is scarcely a district of London which is without its hospital of one kind or another', noted one visitor guide.[7] Many had been established as philanthropic enterprises during the eighteenth century. Guy's, for example, was named after its founder, Thomas Guy, who had endowed the hospital in 1725 with the wealth from his business.[8] Boards of governors controlled hospitals, and Peterson has established the considerable influence they wielded over staff and governor appointments and patient admissions.[9] During this period, there was great tension between the governors who understood the purpose of hospitals to be charitable, and the medical staff whose primary focus was on teaching and extending medical knowledge, particularly with regard to patient selection.[10]

For ambitious surgeons, a hospital appointment was the most effective route to establishing a successful practice. The direct income from such posts was not large; a payment of £50–100 was usual and extra fees were earned by taking on dressers and students. But surgeons gained unparalleled opportunities for networking with other elite doctors and influential governors and establishing a reputation that would win them private patients. Although there had been a substantial increase in the number of junior surgical hospital posts since the 1820s, there were few senior positions.[11] Only 26 surgeon and around 20 assistant surgeon posts existed in the prestigious London teaching hospitals in 1846, whilst in the same year, there were 1000 or so medical students in the city.[12] Thus most would-be hospital surgeons spent long periods waiting for a vacancy to occur whilst they struggled to make a living through teaching and other appointments. It took James Paget 11 years to become assistant surgeon at St Bartholomew's hospital.[13]

For patients, social class was the key determinant of place of care. Poor Law surgeons cared for the very poorest of society in the workhouses, while hospitals treated mainly the working classes. The majority of hospital patients came from parishes local to the institution, although some travelled from the provinces and occasionally from other countries to gain admittance to a particular hospital or to consult a specific surgeon. By 1846, the national and often international reputation of many London physicians and surgeons had established the metropolis as the pinnacle of British medical practice. Patients who could afford it consulted surgeons privately and were treated at home. Those living outside London took lodgings or stayed in hotels: operations were frequently performed in these temporary surroundings. This pattern had long been established amongst elite patients, but the early nineteenth-century growth in the middle-class medical market had brought many new patients to the metropolis.[14]

It is important to recognise that the demography of hospital patients was determined as much by wider social and economic influences as by governor prescription or patient choice. For example, hospitals admitted more male than female surgical patients and this bias can largely be explained through the social organisation of work. More men than women worked in industries which exposed them to accidents from machinery; accident cases formed a substantial proportion of hospital admissions. At the Westminster, the London, St George's and Charing Cross hospitals, for example, accident victims accounted for around half of all admissions and the ratio for surgical patients was around two-thirds male to one-third female.[15] By the late 1840s, most hospitals admitted accident cases at any hour of the day or night, and some like Guy's also offered a daily service for casualties. It was an effective route to increasing medical control over patient selection.[16] Other admissions still required the support of a governor, and were limited to a specified day and time when prospective patients were assessed by the surgeons. Charles Dickens described the out-patients department of St Bartholomew's in 1851:

> The patients enter by the colonnade seen from Smithfield. . . . there are two doors, one for women and one for men . . . by eleven o'clock these apartments are filled with people of all ages, from the baby a

> month old, sickening with measles or hooping-cough, to the old crone of seventy, groaning with old age... The crowd of patients becomes thicker and thicker... [the doctor] commences his first examination of the out-patients – a task that looks enough to occupy the whole day. 'What is it?' is the rapid inquiry; and while these words come from his tongue, his rapid practiced eye is scanning the face of the patient, and his finger is feeling the pulse. The few first words of the patient tell him all he needs.[17]

His description did not exaggerate. Paget assessed between 180 and 220 surgical cases on each of his admitting days.[18] However, the vast majority of patients were treated by therapies or local applications rather than operations.

In the 1840s, surgeons were unequivocal that major operations could be justified only as a means of saving life, and this is borne out in the very low number of operations that were performed.[19] Each London hospital held only one short elective operating list per week – rarely were more than two or three major operations performed – and acute surgery was undertaken only for the most severe trauma cases or conditions such as strangulated hernia. Data from Guy's shows that between January 1845 and March 1846, only two amputations were performed on average per month, and only one or two lithotomies.[20] Liston, who performed the first amputation under ether, recorded only two or three major operations per month in his University College casebooks between 1839–46.[21] At King's College during 1845–46, Fergusson operated on only 15 per cent of male and 8 per cent of female surgical cases.[22] Lesser operations such as amputation of fingers or toes, smaller tumours, cleft palate or harelip repair were undertaken because of their potential to threaten life if left untouched. Damaged tissues and fractured bones left untreated could lead to sepsis. Congenital problems such as harelip could mean that the baby suffered great problems in feeding and thus became emaciated and sickly. Even so, only one harelip operation was recorded at Guy's in 1844; Fergusson performed five at King's between 1845–46.[23] Some minor surgery such as the removal of toenails or small tumours took place in out-patient departments, but any procedure which required the patient to be held down was performed in the operating theatre.

Operative mortality remained high, despite an overall drop in hospital mortality during the first part of the nineteenth century.[24] At St George's hospital between 1842 and 1843, 1 out of 6 patients operated on for hernia died;[25] 4 out of 7 patients operated on for hernia at Guy's died in

1845;[26] and during the same period at University College hospital, mortality for amputations of the thigh, performed on accident cases, was almost 60 per cent.[27] Provincial surgeons reported similar rates. Liverpool surgeon John Halton's statistics of 1843 revealed that 1 patient out of 6 died after a leg amputation or the removal of a tumour, arm amputations resulted in 1 death out of 18 patients, and 1 in 4 patients treated for hernia died. The cause of death in each case was tetanus or gangrene.[28] In view of such figures, it is easy to understand the stress surgeons placed upon assessing the risks and benefits of each operation.

Surgeons rigorously evaluated the risks and propriety of each operation before proposing it to patients. Patients often took time to come to a decision, and they do not appear to have been unduly pressured. 'The patient must decide for himself', said Benjamin Brodie,[29] and Liston confided to his colleague James Miller that he refused to 'try to persuade the patient' as it was too serious a matter.[30] When a 'respectable', married 40-year-old woman presented to St Bartholomew's with a tumour of the vagina in 1846, the surgeon, William Lawrence, proposed to operate. But she was 'unable to make up her mind' so left the hospital, and returned 6 months later ready to undergo the operation.[31] At University College, Richard Quain advised a man admitted with a compound fracture that amputation of his arm was 'the only means of saving his life'. The patient was given time to consult his wife, and the operation did not take place for a couple of weeks. Quain later speculated on whether the arm should have been removed immediately but noted that there 'appeared no certainty of the joint having been injured, and that patients have an extreme and very natural unwillingness to submit to operation'. He concluded that putting the limb into splints and awaiting the progress of events had proved the best course of action.[32]

That patients were aware of their freedom to exercise choice is evident from examples of patients who visited more than one hospital in order to obtain the diagnosis or treatment that they believed was most appropriate. One 4-year-old boy whose mother suspected he had stone in the bladder was taken to St Bartholomew's and Guy's, but no stone was detected. Eventually she took him to King's, where the surgeon, William Fergusson, did detect a stone and performed lithotomy.[33] In 1846 the *Lancet* carried an account of a patient suffering from ulcers on his arm who had been admitted sequentially to Guy's, King's, Charing Cross and the Metropolitan hospitals. Guy's surgeon, Bransby Cooper, had told him that the only chance of saving his life was through the amputation of his arm, but the patient refused to accept this diagnosis.[34]

Although surgeons were frustrated when patients declined proposed operations, they accepted their veto, and it seems that many more operations were offered to patients than were performed. Operations listed in the annual report of the Edinburgh Eye Infirmary of 1836, for example, equalled the number 'recommended but not consented to'.[35] The 1845 and 1846 reports on surgical patients at Guy's specify several instances where patients had refused operations.[36] Thus, gaining consent for operations was a process of negotiation and collaboration, rather than domination: surgeons were bound to demonstrate that their proposal conformed to established surgical principles. When William Lawrence, at St Bartholomew's, suggested that the entire breast of a patient should be removed, he noted that the other hospital surgeons had supported 'the propriety' of his opinion.[37] The same approach was taken within private practice. Benjamin Brodie, past president of the Royal College of Surgeons, called in William Fergusson to advise on a prospective lithotomy case.[38] Elite surgeons were also used as sounding boards for those beyond the metropolis. In 1846, James Luke, surgeon at the London hospital, was consulted by a surgeon from Woodford Green on 'the propriety of trephining over the seat of a fracture'.[39]

But despite the small numbers of operations, and the rhetoric of conservative practice, the operating theatres of the London hospitals remained the key arena for displays of surgical prowess. Many hospitals admitted members of the public to watch operations, and there were always large crowds of medical students and other doctors present. By the late 1840s, 'modern' surgeons had constructed their professional identity upon attributes such as coolness and decisiveness. It was an image with elements of showmanship; the surgeon was the oasis of authority amidst the bodily confusion of severed flesh and bones, and the disarray of minds. The surgical focus was upon the body: ensuring the body was in the most suitable condition possible, deciding on the procedure, the most appropriate incisions and wound closures. These were the sorts of criteria on which his peers would judge a surgeon: no London surgeon was unaware of the way in which their public performance in the operating theatre could make or mar private practice.[40] These surgical pressures were exacerbated by the pain and suffering of patients during operations. And, as Chapter 1 suggested, by 1846, from all perspectives, the pain of surgery had become an acute problem.

## The landscape of oblivion

Within a month of the news of its 'invention' arriving in London, ether had been trialled in every teaching hospital during major operations such as amputation, lithotomy and hernia, as well as in the removal of tumours, toenails and teeth. Aside from the obvious humanitarian benefits of insensibility to pain, when ether worked well, it could transform surgical efficiency. At St Bartholomew's, Lawrence removed a diseased eye from the orbit of a middle-aged man and marvelled at the improvement ether brought. The patient slept well on the night prior to the operation as his 'mind was tranquillised by the belief that the new process would lessen or prevent pain'; the next day, the patient 'lay like a body on the dissecting table'; on waking, he remembered nothing of the process and had felt no sensations, and he recovered well with very little subsequent pain. Lawrence compared these events to a previous excision when the patient had suffered intensely, and 'writhed in agony, not being able to control himself'.[41] He also noted that the insensibility produced by ether had saved 20 minutes of the operating time.[42]

Ether relaxed muscles and made it possible to reduce longstanding joint dislocations.[43] A strangulated hernia could be remedied without operation, and Liston inserted a catheter in the bladder of one patient under ether without having to cut the perinaeum.[44] Children responded particularly well to its effects, and some of the first patients to be operated on were a little boy at St Bartholomew's who had ruptured his urethra by a fall, and a 6-year old at St Thomas': 'even when the hand was held before him, the child would not believe that the finger had been removed'.[45] They were amongst 'the most favourable' subjects for ether, said Snow. Not only did they recover quickly from its effects but it prevented their 'struggles' which would often interfere with the performance of an operation.[46] Yet despite these strong advantages, the use of ether created many practical problems beyond concerns about physiological or moral risks. A body without feeling was also a body without control.

Prior to ether the surgeon and patient functioned as a symbiotic unit. Patient cooperation and complicity were vital factors and contributed to the success of the procedure. 'A surgeon must be well assisted by the patient, or he cannot succeed', Liston had told his medical students.[47] But ether removed the patient's capacity for control. At times it caused such rigidity in muscles that patients had to be physically restrained – as they had prior to ether. Operations that were usually performed with the patient seated in a chair became problematic; muscle rigours could jerk the body out of its position and complete insensibility caused it to slump forwards. At St George's, ether caused one woman to struggle so much that the surgeon, Caesar Hawkins, refused to begin the operation – the removal of a large ovarian tumour. As the patient would not allow the operation without pain-relief it was aborted.[48] Lithotomy and lithotrity were usually performed with the patient lying on their front, or bending forwards over the table – not easy positions in which to inhale gas. Snow's solution was to place the patient on their side with their knees drawn up, but not all surgeons were prepared to change their practice.

Even when control was not an issue, ether could complicate surgery. At King's College after a patient had been operated on for fistula without ether, the notes record that: 'the operation was accomplished with much greater facility than in two cases where the patients [were] rendered insensible... ether... [produces] many obstacles to the skilful performance of an operation'.[49] So although ether resolved the inherent humanitarian conflict of surgery – causing suffering in order to preserve life – it also created new problems. Nevertheless, it was brought into regular use at the weekly operating sessions in many London hospitals.[50]

The availability of pain-relief changed surgical practice from the beginning. Analysis of around 70 operations performed throughout London during the first 3 months of 1847 shows that ether reconfigured the spread of operations – Appendix Tables A.1, A.2, A.3. The demographic patterns of surgical patients had not changed – the majority were aged between 21 and 49 years, and twice as many men received operations as women – but a shift can be discerned in the type of operations performed. These included several excisions of joints – acknowledged to be one of the most painful procedures – and the number of major amputations was equalled by minor procedures such as toenail removal, lacerations and circumcision. Such conditions were not of immediate threat to life and surgeons had been willing to undertake them prior to ether to prevent the development of serious infection. But the prospect of pain had caused many patients to refuse. As a consequence, cases were frequently left to deteriorate to the point when amputating the limb was the only means of saving life. Ether immediately began to reverse these trends: the promise of insensibility caused many patients to submit to early intervention. At the Middlesex hospital, a 17-year-old youth with ulcerated skin surrounding both his great toenails agreed to their removal if it could be done without pain, and a countryman who had damaged his finger in a chaffcutting machine was persuaded to have it amputated immediately.[51] An Irishman was admitted to the

London hospital with a compound fracture and dislocation of his leg, caused by a cask of sugar falling on him whilst he was working in St Katherine's Docks; he refused to submit to an amputation until he was told it could be done without pain.[52]

That ether's powers were widely known from the beginning is evident from the many examples of patients who presented themselves at hospitals requesting to use the new pain-relief.[53] Chapter 2 showed how the first patients to try ether at King's College were those who had previously refused operations on account of the pain. This shift in patient attitude seems to have produced a surge of operations on cases of long-standing disease. At St George's, for example, over 90 per cent of arm or leg amputations performed under ether during the first 5 months of 1847 were on account of chronic disease of the ankle or wrist joint.[54]

By April 1847, the number of surgical operations in London had 'more than doubled', noted the *Lancet*.[55] The increase in surgery fuelled a concern that many unnecessary operations had been performed under ether. But although the number of operations did increase during 1847, there is no evidence that this arose because of surgeons revising their calculus. Rather it reflected the strong and positive response of patients to painless surgery. Surgical notes reveal no difference in the treatment of patients admitted to the hospitals after December 1846. Martha King spent 2 months in King's College in 1847 whilst various poultices and bandaging procedures were used to try and avoid the amputation of her leg, and 8-year-old Jane Twip spent 5 weeks on the ward before the decision was taken to operate on her diseased ankle joint.[56] When a 31-year-old man presented to St George's with a dislocated shoulder in March 1847, it was 10 weeks before the joint was reduced under ether.[57] And surgeons continued to classify patients as inoperable, due to either the location or severity of the disease. A 66-year-old Irish engineer who presented to King's College with a stomach tumour was sent home as 'nothing can be done... in the shape of an operation'.[58]

We get a more detailed view of these early shifts by comparing the surgical practice of King's College surgeon, Fergusson, pre- and post-ether. Fergusson trained in Edinburgh and became Professor of Surgery at King's College and surgeon of its hospital in 1840. By 1846 he was acknowledged to be one of the most talented surgeons in the metropolis. He had developed a strong interest in the treatment of fractures by conservative methods and was particularly adept in performing cleft palate and harelip operations.[59] Between 1845–46 and 1847 the number of overall hospital admissions increased slightly – around 7 per cent – and out-patient numbers rose by 8 per cent (Figure 5.1).

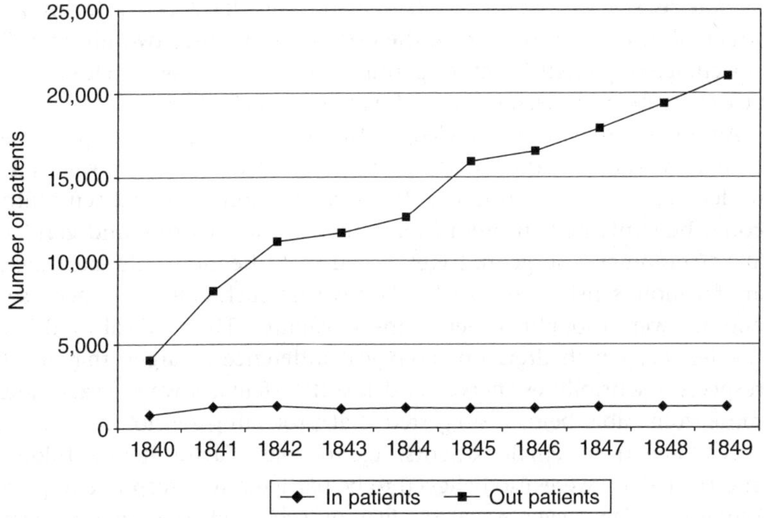


*Note*: Both medical and surgical patients are included.

*Figure 5.1* Patients treated at King's College Hospital, 1840–49.
*Source*: King's College Hospital annual reports.

But Fergusson's operating seems to have doubled. Whereas before ether he operated only on 1 in 8 patients, after ether, 1 in 3 patients received operations – Appendix Table A.4. That the rise was due to ether is also suggested by the spread of operations. In 1845–46 Fergusson had performed two excisions to remove dead bone from limbs which accounted for 12 per cent of his total operating; within the first 8 months of ether he performed seven similar operations which accounted for 20 per cent of total practice – Appendix Table A.5. Fergusson did not use ether universally: only 1 in 3 patients in this 1847 sample received it. But its use in all operations on bones or joints suggests that it was the pain of these procedures which had previously discouraged their performance. Fergusson's particular skills in harelip and cleft palate work meant that his surgical practice consisted of a comparatively high proportion of reconstructive work prior to ether. Although the proportion of reconstructive work rose slightly between 1845–46 and 1847, ether was not used for many of the procedures because of the difficulties in combining the face-piece of the inhaler with operations on the mouth and nose, particularly for harelip where the risks of haemorrhage made it incumbent upon the surgeon to operate as swiftly as possible. (Fergusson had developed a technique which allowed him to perform the operation in a bare two minutes.[60]) His choice of patient for ether permits some interesting conclusions to be drawn about understandings of pain and sensibilities.

Although surgeons acknowledged that patient tolerance to pain was extremely variable, the broader influences of age and sex were also understood to affect sensibility. It was a common view that sensibility could be explained through hierarchies of race, culture and gender. St Bartholomew's surgeon, Paget, noted that 'the more cultivated races are far more sensitive to pain'.[61] And within such cultivated races, sex and age were thought to determine sensibility. The medical model of the female, which drew on biological difference to argue that in all respects – sensibilities, nerves and health – females were weaker and more vulnerable beings, suggested that protecting sensibilities was of special benefit to female patients. Age too was a factor, hence children and the elderly were also believed to be particularly susceptible to pain. Fergusson's data suggests that he drew on such understandings to determine his choice of patient for ether. Between February and August 1847, patients aged between 21–50 years were least likely to benefit from ether; only 44 per cent of those patients experienced painfree surgery, compared to 50 per cent of under 10s, and 78 per cent of over 50s. And whereas 44 per cent of female patients received ether, only 33 per cent of males did so – Appendix Tables A.6, A.7, A.8.

The data on Fergusson's practice is remarkably similar to the patterns established by Pernick in his analysis of the effects of ether on surgical practice at the Massachusetts General Hospital. There, the rate of surgery increased by 2.5 times during the first 12 months of ether use and the patients most likely to receive ether were those understood to be particularly vulnerable to pain: women, children and accident victims.[62] Fergusson's rate of surgery increased by 2.6 times, and his selection of patients met similar criteria. Thus it seems that despite the different methods of administration that we noted in Chapter 2, ether was given to a similar selection of 'vulnerable' patients in London and Boston, and surgery increased at a similar rate.

The introduction of chloroform in November 1847 sustained the growth established by ether, and the number of operations performed under anaesthesia in London continued to rise. Chapter 2 showed how keenly surgeons welcomed chloroform after the difficulties of ether. Ether remained in use as a stimulant and was often given to patients recovering from chloroform anaesthesia, but it was rarely used as an anaesthetic in London practice. The surgical records analysed during the course of this study carry few references to ether anaesthesia after 1847; Snow used it in only 0.3 per cent of administrations recorded in his casebooks. Nevertheless, for all the ease of chloroform, the knowledge that the gas could kill without warning caused anaesthesia to remain a selective practice throughout the 1850s. There is strong evidence that most surgeons were unnerved by fatalities and responded by tempering their use of chloroform.

The first death in a London teaching hospital occurred at St Thomas' in 1849, then, during June 1850, a patient died under chloroform at Guy's whilst the surgeon, Cock, was removing part of a diseased hand. The cautious treatment of a 16-year-old boy with a strangulated hernia, later that same year, suggests that the death had caused surgeons to be extremely prudent in the use of chloroform. Alfred Poland first attempted to reduce the hernia using warm baths and manipulation, but the patient's 'intense agony' forced him to desist. It was only at this point that the patient was made insensible with chloroform.[63] Snow noted that the surgeons of both St Thomas' and Guy's were strongly opposed to chloroform on account of its risks, and used it infrequently during the first few years. It seems likely then that these early fatalities had established a wariness of chloroform within those hospitals. (Although Snow believed that this strategy exacerbated rather than contained risk because it failed to establish familiarity with administering chloroform.[64])

Variation in surgical attitudes to the risks of anaesthesia produced a wide variety of experiences for patients, even within the same hospital. Governing boards could create a supportive environment for the take-up of anaesthesia by granting funds for drugs, equipment and staff, but the decision on its use in each operation was taken by the surgeon.[65] A sample of operations performed at St Bartholomew's during 1850 is a good example of this apparently arbitrary approach. During the year, 26 operations took place and chloroform was used in 46 per cent of the total cases. Edward Stanley used it in 67 per cent of his operations, William Lawrence in only 36 per cent of procedures, and Arthur Lloyd did not use it at all.[66] Stanley used pain-relief during an amputation; Lawrence and Lloyd did not. Stanley used it for the majority of hernia cases, but not for lip operations. And although he warned one patient of the possible adverse consequences of having a foreign body extracted from the knee, and spoke of 'the severity of the operation', he did not give chloroform. Lawrence's patients received chloroform whilst cysts were removed from the larynx and the shoulder, but not during lithotomy or phymosis. Lloyd did not use chloroform in any of his three operations: a strangulated hernia, a wound on the forearm and the amputation of a leg.[67] In retrospect, it is easy to judge this sort of practice as random and disparate. Yet it grew out of a conviction that anaesthesia was not a predictable process; each inhalation had the potential to end in death. Surgical attention therefore remained focused on the preservation of life, even if this resulted in operations continuing to cause suffering and pain. The way in which perceptions of risk determined practice is most vividly evidenced by the dip in the London uptake of anaesthesia during 1854 – the same period in which seven fatalities occurred in the metropolis.

## A climate of risk c.1854

The risks of chloroform were acknowledged from its introduction, and Chapter 4 showed the wide publicity that was given between 1850 and 1851 to its potent and threatening powers in both the public and medical domains. Nevertheless, hospital records and Snow's casebooks show that the numbers of operations performed under anaesthesia increased consistently until around 1853–54 when there was a significant dip in the numbers of operations carried out under anaesthetic. At the London hospital, the operations logbook shows that in 1852, 60 per cent of amputations were performed under chloroform. In 1853 this proportion fell to 14 per cent, and during 1854 only 9 per cent of amputees were given anaesthesia (Figure 5.2). At King's College, the number of operations performed in 1854 under anaesthesia fell by 18 per cent from 1853 levels (Figure 5.3). And during the same year, the number of surgical anaesthetics administered by Snow dropped by 13 per cent (Figure 5.4).

The decrease in the London use of chloroform was a local response to a local concern; seven fatalities had occurred in hospitals throughout the metropolis between March 1853 and December 1854 – Appendix Tables A.9, A.10. There is little doubt that the occurrence of so many deaths under the hands of the most experienced surgeons had heightened awareness to such a pitch that even the most confident surgeon curtailed anaesthetic use where possible. In 1852, Fergusson had spoken of anaesthesia as a 'custom' which 'has rapidly gained an undoubted and sure position'.[68] Records confirm his use of chloroform to be near universal by this point. Yet when he performed lithotrity – acknowledged to be a very painful process – on a 51-year-old male in January 1855, he did not sanction the use of chloroform; there was 'very little irritability of the bladder and urethra' he said. Snow, who was in the King's College operating theatre, monitored the patient's pulse during the

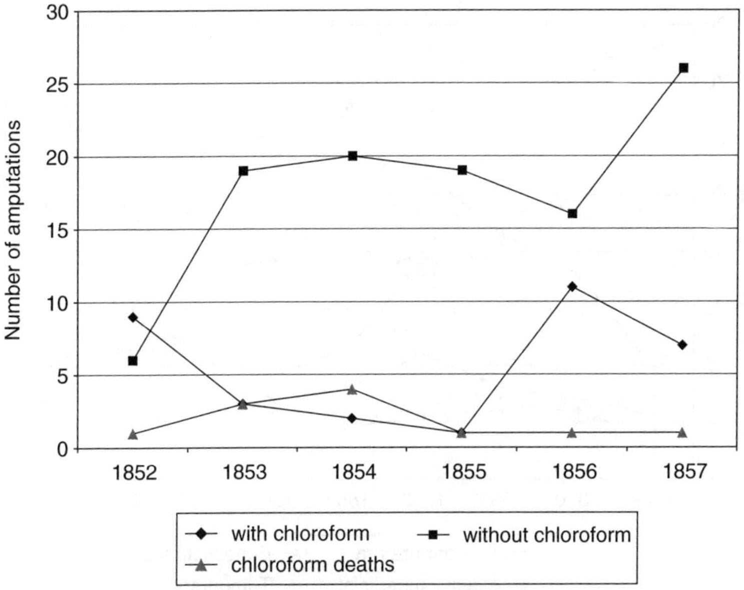


*Note*: Deaths are for the city of London, surgical data is for the London Hospital only.

*Figure 5.2* Amputations at the London Hospital, 1852–57.
*Source*: LH/M/3/74, London Hospital.

operation. As soon as the lithotrite was introduced, the pulse rose from 120 to 144 'and immediately afterwards it became uneven, irregular and intermitting. I could not count more than 3 or 4 beats at a time and occasionally, when the pain seemed greatest, the pulse was altogether imperceptible for about 5 seconds at a time'. The patient was grasping the table so firmly that Snow wondered if this accounted for the pulse pattern. He put his ear against the patient's chest, and noted that 'whilst the large catheter for injecting the bladder was being introduced and the patient was holding his breath, there was no sound whatsoever within the chest'.[69] Fergusson regularly employed chloroform during lithotrity and according to Snow, the patient was 'very healthy'. But only a few weeks previously, a woman had died under chloroform at Guy's, during the amputation of her leg – the seventh fatality in London in only 18 months. Fergusson may have been responding to chloroform deaths, yet from Snow's perspective, the response of the pulse to the pain of the operation highlighted the risks of surgery without insensibility. (Throughout the period, Snow's attitude to anaesthesia remained consistent – any patient fit for

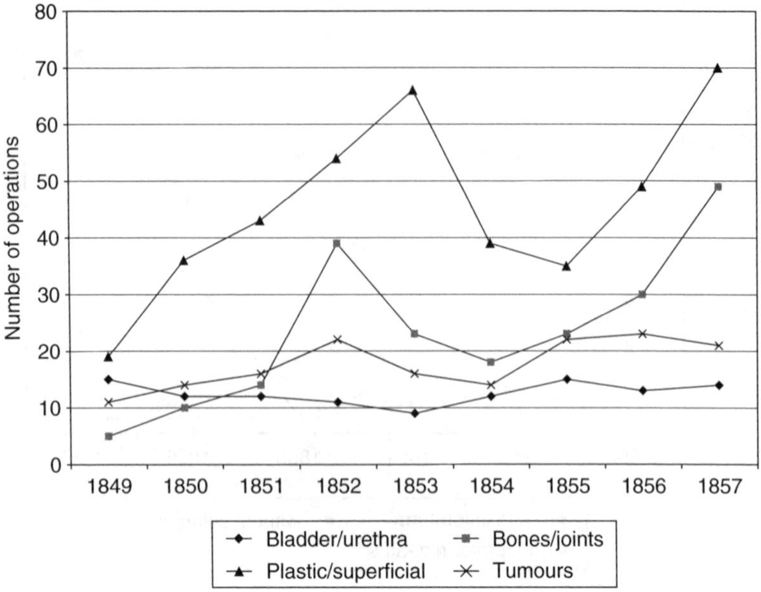


*Figure 5.3* Operations under anaesthesia, King's College Hospital, 1848–57.
*Sources*: KH/CN/71–83, *CB*.

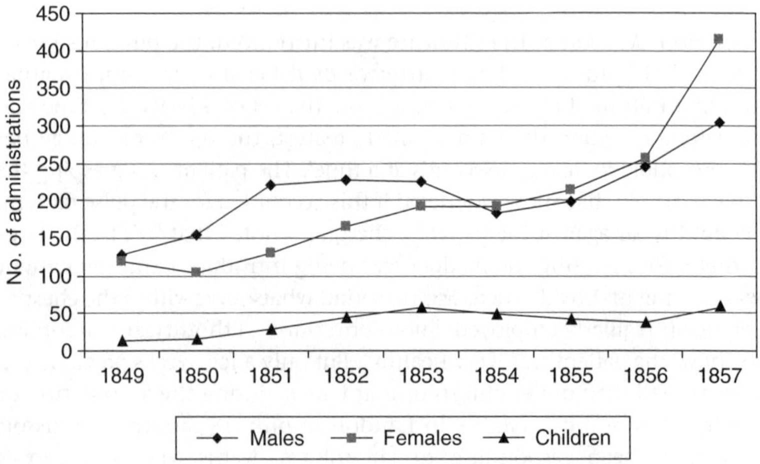


*Figure 5.4* Snow's anaesthetic administrations, by sex, 1849–57.
*Source*: *CB*.

surgery was fit for anaesthesia – and he stated several times that he had never refused to anaesthetise a patient.[70])

That it was surgeons who curtailed the use of anaesthesia during 1854, rather than patients who were too fearful to inhale gas, is borne out by the way in which Snow's hospital anaesthetic practice declined more sharply than his private practice. The number of hospital patients he treated in 1854 fell by 17 per cent from 1853 levels, whereas private anaesthetics dropped by only 6 per cent (Figure 5.5). Breaking down these overall decreases in each sector by the type of procedure performed illuminates the different attitudes of surgeons and patients towards the risks of anaesthesia.

The numbers of hospital anaesthetics administered by Snow fell across all types of procedures – amputations, bones/joints, plastic/superficial, tumours – except for the operations of lithotomy and stricture, where numbers increased from 8 patients in 1853 to 12 in 1854 (Figure 5.6).

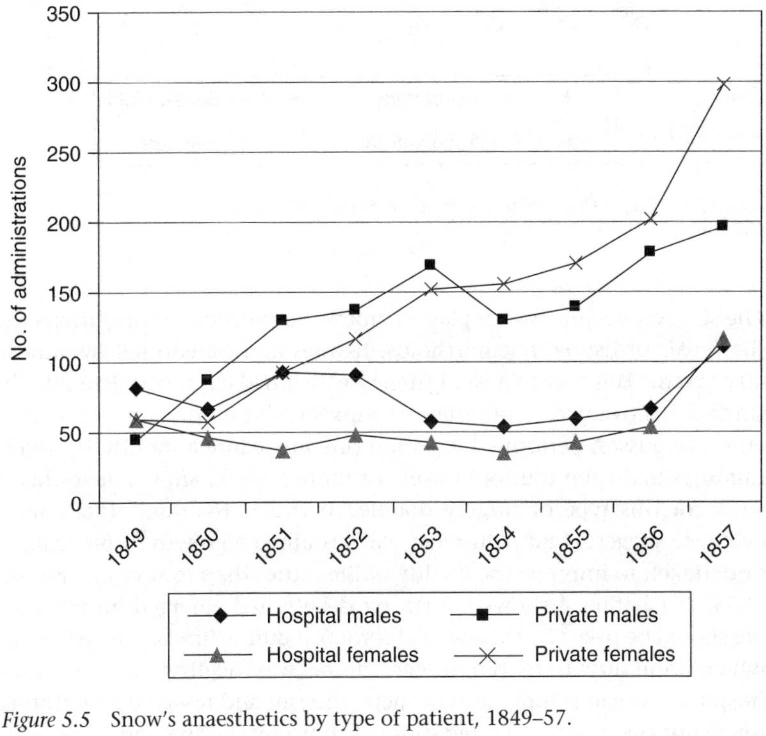


*Figure 5.5* Snow's anaesthetics by type of patient, 1849–57.
*Source*: *CB*.

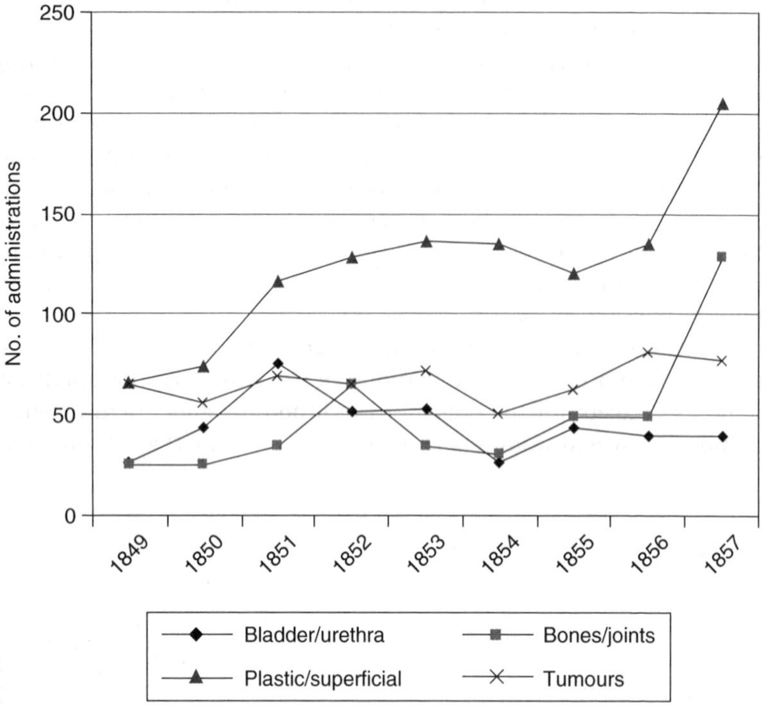


*Figure 5.6* Snow's anaesthetics, by procedure, 1849–57.
*Source*: *CB*.

The steepest decline was in plastic/superficial procedures; operations for the repair of fistula, haemorrhoids, fissures and harelip fell by almost 40 per cent. This was an area of surgery which had expanded dramatically since the introduction of ether. At King's College, for example, before ether, Fergusson performed less than one procedure a month, by 1853, numbers had risen fourfold to one or more a week. Snow's anaesthetic work for this type of surgery doubled between 1849 and 1852. Such evidence suggests that pain-relief had sustained a growth in procedures undertaken to improve the quality of life rather than to save life. But by 1854, as Chapter 4 showed, a strong debate had emerged on whether life should be risked by the use of chloroform during 'trivial' procedures in which mortality from the surgery alone was negligible. In London hospitals, surgeons took heed of such criticism and reserved anaesthesia for major operations – amputations, lithotomy, hernia and so on – in which the benefits of obviating the pain and shock of the operation were believed to outweigh the risks of chloroform.

It also seems that surgeons' awareness of anaesthetic risk overrode concern for the susceptibility of hospital women and children to pain. During 1854, the numbers of hospital female patients to whom Snow gave anaesthetis dropped by 14 per cent; those of infants by 39 per cent. Anaesthetics given to hospital male patients, however, fell by only 7 per cent (Figure 5.5). In part, this difference was caused by the established bias towards male surgical patients, many of whom were accident victims requiring major surgery. In 1854 at the London hospital for instance, 20 male patients received amputations compared to 2 females,[71] and out of seven hospital amputations Snow attended that year, only one patient was female. That more hospital male patients underwent serious surgery than females is also suggested by the increase in anaesthetics for hospital patients undergoing lithotomy or stricture during 1854, the majority of which were male. But the details of Snow's private anaesthetics reveal a different story.

During 1854, although Snow gave fewer anaesthetics to private patients for major procedures such as amputations, bones/joints and tumours, other areas increased significantly. Anaesthetics for plastic/superficial operations rose by 32 per cent, those for eye operations by 30 per cent, and those for dental work by 24 per cent (Figure 5.7). Surgeons classified such procedures as low-risk and in hospital practice had demonstrated a reluctance to increase danger by using anaesthesia. Nevertheless, the fact that the use of anaesthesia in private practice continued to thrive during the crisis period of 1854 suggests that from a patient's point of view, the fear of pain – even for a tooth extraction – was stronger than the fear of anaesthetic death.

It is remarkable that during this time of heightened awareness of risk, when surgeons were clearly curtailing their use of anaesthesia in the hospitals, there is no evidence of a return to ether in London practice. As we saw in Chapter 2, the occurrence of chloroform fatalities during 1848 and 1849 had caused the northern states of America, and parts of Europe, to return to ether by the early 1850s. American surgeons had developed an efficacious method of administering ether so it was perhaps no hardship for them to abandon chloroform. It is likely that the growing incidence of medical malpractice suits in America caused surgeons to be particularly alert to risk and be prepared to tailor their practice accordingly.[72] But the British dislike of ether persisted and surgeons continued to give priority to the ease of chloroform despite the fact that their fear of its risks caused anaesthesia to remain a selective

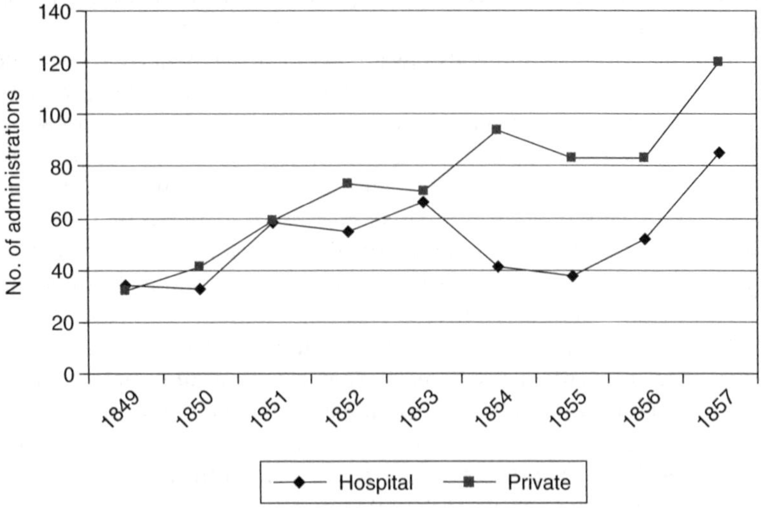


*Figure 5.7* Snow's anaesthetics for plastic/superficial operations, by type of patient, 1849–57.
*Source*: *CB*.

practice throughout the 1850s. Nor indeed did they pursue other methods of inducing insensibility such as the surgeon James Arnott's proposal that the skin at the site of the operation could be temporarily frozen so as to numb sensation.[73]

By 1856, the London crisis of confidence in anaesthesia was over. Both at King's College and in Snow's practice the use of anaesthesia had risen to earlier levels and was continuing to increase. In part, this return of confidence had been fuelled by the successful use of chloroform during the Crimean war, despite the initial warning of Sir John Hall, principal medical officer, that army medical officers should be exceedingly cautious about its employment. It is useful to examine the detail of Hall's arguments as they reveal the rationale behind the selective use of anaesthesia which was as much part of civilian practice as that of the battlefield during this period.

## Shock and chloroform

In September 1854, within days of allied troops landing in the Crimea, John Hall issued a memorandum for army medical officers which cautioned against the use of chloroform for amputations following the 'severe shock' of gunshot wounds. The 'smart of the knife is a powerful stimulant' he warned, and continued: 'it is much better to hear a man bawl lustily than to see him sink silently into the grave'.[74] The timing of Hall's memorandum, coinciding as it did with the overall dip in the use of anaesthesia in London practice, suggests that he was influenced by the pervading fear of fatalities. He guessed that public opinion, founded on 'mistaken philanthropy', would be against him. And, indeed, the publication of his memorandum in the *Illustrated London News* provoked a 'storm of abuse, indignation and misrepresentation'.[75] The conflict stemmed from the different attributions given to pain. For Hall and many others, pain performed an essential function, and the insensibility produced by chloroform exacerbated the risks of surgery. Indeed, the death of a patient under chloroform at University College hospital during the dispute led one correspondent to write to the *Times* citing the fatality as a clear justification of Hall's prudent approach.[76] Amongst those to challenge Hall's views was Edinburgh surgeon, James Syme. He repeated Snow's explanation that rather than being a powerful stimulant, pain exacerbated the effects of shock on an already weak patient; pain was not simply purposeless, it contributed a significant physiological risk.[77] These arguments were taken up in many quarters of the public and medical press but despite the outcry provoked by Hall's memorandum, it is clear that such views were not unusual amongst surgeons, and throughout the 1850s the use of chloroform in amputations remained selective.

Between 1852 and 1857, for example, 124 amputations were performed at the London hospital, yet only 28 per cent of patients were given chloroform (Figure 5.2). There is no doubt that the injuries of such patients were appalling and they suffered the highest mortality rates: only one out of three railway guards who sustained compound fractures to their legs during 1852 survived. Yet not one of these patients was given chloroform before surgery, which in two cases was performed by Thomas Curling, lecturer in surgery at the hospital. A few years earlier, Curling had addressed the Hunterian Society on the advantages of ether and chloroform in operative surgery and spoke of 'the ill consequences arising from exhaustion of the vital powers' in operations, and of the ability of anaesthesia to exert a beneficial influence on accident patients.[78] Why therefore did he not use chloroform? To answer this question we need to return to the understandings of anaesthesia discussed in Chapter 4 – was it a therapy or a process?

It was acknowledged that victims of gunshot wounds and the most serious industrial or railway accidents were often in a severe state of shock. For these patients, the only surgical recourse lay in the amputation of mangled limbs as a means of stemming blood loss and of removing potential sites of sepsis. But it was not uncommon for the most severely injured patients to be deemed unfit for surgery at all. In such cases, surgeons believed the shock of the injuries had depressed the patient's nervous system to a point where it could not withstand surgery. That a patient had to be 'fit for surgery' was a long established principle; the shock of an operation was understood to depress the body's systems and although the pain of the knife was a counter stimulant, patients had to be sufficiently strong to begin with. For this reason, the use of stimulants like alcohol, both before, occasionally during, and after surgery was common practice. In the cases at the London hospital, at least one of the railway guards was rallied with brandy and ammonia in order to undergo his operation. From this perspective, administering a depressant in the form of chloroform would have been constructed as exacerbating, rather than reducing surgical risks. Other cases support this interpretation: a 40-year-old married woman who presented with a strangulated hernia was specifically not given chloroform because 'she was so low', nor was a 61-year-old widow who had her thigh amputated after a chronic ulcer, because of concern that she was already in a 'very unfit state' for surgery.[79] During this period then, most surgeons were determining their use of chloroform within the longstanding therapeutic dichotomy of stimulants and depressants which read the body as a holistic system. Hall's memorandum expressed this view. Nevertheless, in practice, chloroform was widely used by army officers during the Crimean war.

A series of returns of wounds and injuries admitted to army hospitals in the Crimea and covering the period from September 1854 to December 1855 shows that 60–95 per cent of amputations were performed under chloroform.[80] Confirmation of this shift of opinion in favour of chloroform can be found in the Government review of medical services during the Crimean war, published in 1857. The majority of surgeons supported the use of chloroform in 'both severe and slight wounds requiring operations', although, it concluded, some remained unconvinced that its risks were justified in minor operations.[81] There is no evidence that ether was used as an anaesthetic during the Crimean war.[82] Aside from its propensity to kill without warning, chloroform had natural advantages over ether that made it particularly suitable for the battlefield; it was not inflammable like ether, and most individuals could master the technique of administering it on a cloth.

The reconfiguration of chloroform as a benefit rather than a risk for severely shocked patients can also be discerned in the management of accident cases. At St Mary's hospital, in 1862, a railway porter was admitted after a 36-ton engine had run over his legs. Death seemed imminent and when his clothes were removed 'a shocking spectacle presented itself: the left thigh was literally smashed; the muscles were lacerated and exposed, with the vessels and nerves…haemorrhage which was taking place could not be checked by pressure upon the femoral artery'. Surgeons decided to remove both limbs immediately under chloroform. His condition was so extreme, they did not attempt to move him to the operating theatre, but performed surgery in the ward. Afterwards, the patient was able to take some milk and brandy, but subsequently died. Although the case had been regarded as 'almost hopeless' from the beginning, 'the operation was resorted to for humanity's sake, and as the only means of saving the poor fellow's life'.[83]

Twenty, and possibly even 10 years earlier, the patient would have died from severe blood loss in a state of acute pain; surgeons would have accepted they had little to offer such individuals. But anaesthesia had changed surgical practice. Although the majority of surgeons did not initially share Snow's view that anaesthesia decreased the physiological risks of surgery, by the late 1850s the balance of opinion had changed, and 'shocked' patients were more likely to be anaesthetised before surgery. It suggests that the view of anaesthesia as a process, which could produce insensibility without interfering with circulation or respiration, was becoming more common. In army and civilian hospitals the practice of selective anaesthesia could be sustained because surgeons held jurisdiction over its use. In private practice, however, the different dynamics of the patient–surgeon relationship created a different balance of power.

## The powers of private patients

Within hospital walls, surgeons were implacable on withholding anaesthesia if they believed the risks were too great. Hospital patients could refuse treatment but the surgeon's clinical judgement was the key determinant of practice. In private practice though, a mixture of economics and etiquette informed the relations between surgeon and patient. Visiting and treating patients in the home environment meant that surgeons were bound to conform to social proprieties.[84] Operations were performed in sitting rooms or bedrooms, with patients sitting in a chair or lying on a bed, attended by their family and friends.

In an overcrowded London market, few could afford to ignore the fact that patients had a propensity to change doctors if they were dissatisfied with their care: Snow gave anaesthesia to many individuals whose regular doctor would not condone its use.[85] Meeting the patient's expectations then was of prime concern, but it was not always easy to unite this objective with clinical principles. The following example shows one incident in which Snow almost caused a chloroform fatality because of attempting to subdue the cries of an infant undergoing surgery.

In June 1853 he attended a 'rickety and diminutive' 6-year-old child having part of its eyeball removed. When insensibility was established, the inhaler was withdrawn to allow Benjamin Travers space to operate. (In cases where the inhaler interfered with the surgery Snow used a sponge to maintain anaesthesia.) As the needle entered the skin, the child gave a cry. Snow, knowing the parents were in an adjoining room, responded instantly:

> I poured some chloroform, hastily and without measuring it, on a rather large sponge and placed it over the nostrils and mouth. It became pressed down upon the nose (more) than I intended by Mr Travers' hand, but was removed after the child had taken a few inspirations. The operation was quickly concluded without any further signs of feeling. At the conclusion of the operation the breathing was natural but the face was pale and the lips blue, and the limbs were also relaxed. I tried to feel the pulse at the wrist but did not discover any. However, I did not at first feel uneasy as the breathing was going on well and the chloroform had been left off some time... in a little time, the breathing became slow and embarrassed and then appeared about to cease altogether, the pulse still being absent. The windows were opened and cold water dashed freely on the child's face.... In a moment or two the child was red in the face and crying violently from pain.[86]

When he later wrote about this incident, he acknowledged that disaster had only narrowly been avoided:

> There is no doubt that in this case the heart was paralysed, or nearly so, by the chloroform, and that its action was restored by the spontaneous gasping inspirations of the child. The accident could have been prevented by having the chloroform, which was put on the sponge, diluted with spirit.[87]

He candidly admitted that his hastiness with the sponge was prompted by the thought of the parents becoming 'alarmed', and this instant response caused him to compromise his principles for the safe practice of anaesthesia.

Surgeons too, like Snow, often blurred their clinical principles within the context of private practice in order to satisfy patients. For instance, Frederick Salmon, who specialised in rectal surgery and established the Fistula Infirmary, later St Mark's Hospital, was reluctant to use anaesthesia because of the risks of bleeding.[88] However, Snow's casebooks record a total of 118 private operations under chloroform which were carried out by Salmon between 1852 and 1858. During these operations Salmon practised the same procedures as in the operating theatre at St Mark's; he treated fistulas, fissures, strictures and haemorrhoids, and performed preliminary examinations to assess the extent of disease.[89] Snow makes no mention of any untoward bleeding in his accounts of these operations, although it is clear that Salmon's arrogant and demanding personality created tension in the relationship. More than once, he refers to the difficulty of 'getting the patient to keep as quiet under the operation as Mr Salmon wished'.[90] Salmon was prepared to sanction anaesthesia, albeit in an antagonistic manner, for his private patients, although apparently not for those he treated in the hospitals.[91] We cannot seek to form a retrospective judgement on the veracity of such medical judgements; on whether the perceived risks of the individual case were justified; but it is clear that private patients had considerable power to negotiate their use of anaesthesia.

There is little doubt that private patients – especially women – used the advantages of their position to capitalise upon the benefits of painless surgery and dentistry (Figures 5.5 and 5.8). During the period, despite the well-established risks of chloroform fatalities and the strong criticism of its use in 'trivial' operations, the number of anaesthetics Snow administered to private female patients for plastic/superficial procedures increased fourfold, and dental work increased by 17 times. Ophthalmic procedures grew in number too. When chloroform was first introduced, many surgeons were reluctant to employ it for eye surgery because they feared that post-operative vomiting – a common side-effect of chloroform – would exert pressure on the eyeball and prejudice the outcome. Thus in 1849, Snow only recorded one eye operation. But by 1857, surgeons had gained confidence and Snow was giving anaesthesia to around two female eye patients every month. That painless surgery encouraged patients to submit to procedures that enhanced life or improved looks is suggested by the fact that around 60 per cent of eye operations during

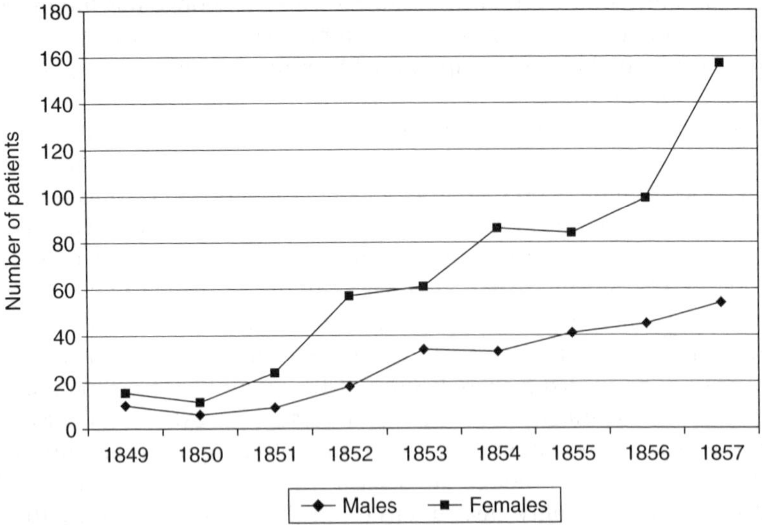


*Figure 5.8* Snow's dental anaesthetics, by sex, 1849–57.
*Source*: *CB*.

1857 on private female patients were to correct squints. Snow's practice of dental anaesthesia is another strong example of this shift.

## Perfect teeth

On learning of the discovery of ether, Charlotte Brontë's immediate thought was that her front teeth could be 'extracted and rearranged' without pain.[92] By 1846, the evolvement of dental technology had improved the fit and appearance of artificial sets of teeth. Wax was used to take impressions from patients' mouths and these impressions were then used to make gold or platinum base plates. Porcelain teeth replaced those made from gold, ivory or wood.[93] But, as Chapter 1 showed, dentists could not palliate the pain of extractions. In the same way that it deterred patients from surgery, it would also seem that the prospect of pain inhibited dentistry. That the growth in Snow's dental anaesthetic work was largely sustained by middle-class, middle-aged women links nicely to the parallel female uptake of surgery. And the story of one of Snow's most strong-minded patients shows how for many women, painless dentistry became a quest.

In January 1849, Mrs Charsley decided upon the use of chloroform for the extraction of some teeth, but her 'medical man in the country', strongly opposed its use on account of a heart irregularity. She travelled to consult King's College surgeon, Richard Partridge. He regularly operated under chloroform but was reluctant to form an opinion until she had seen Snow. Mrs Charsley lost patience and took herself to the dentist, Robinson, who had performed the first tooth extraction under ether in 1846. He gave her chloroform, but despite his wide experience, he failed to establish insensibility. She departed Robinson's surgery in Gower Street and went directly to Snow. He examined her, found an irregularity in her pulse which he believed was not significant, and returned with her to Robinson's. Together, Snow and Robinson began to give chloroform. As Mrs Charsley inhaled, she pushed away the face-piece and said she could not go on: 'she felt as if she was going to die'. After several attempts, the face-piece was abandoned, and a hanky saturated in chloroform was wafted in front of her face. She became drowsy enough for the face-piece to be replaced; she was subsequently made fully insensible and the teeth were extracted.[94]

Her story reveals a remarkable persistence: although her fear of chloroform was strong, her fear of pain was stronger still. It also shows that, though she fell foul of the etiquette which was supposed to determine medical consultations – ignored the advice of her country doctor, failed to follow Partridge's advice, turned up on Snow's doorstep without medical introduction – her wishes were complied with without question. Robinson and Snow were prepared to assist her in every way to achieve her goal of painless tooth extraction. Mrs Charsley may have been extreme in her flouting of doctor–patient etiquette, but she was not unique in her desire for painless dentistry. For those women (and men) who could afford to pay, anaesthesia provided the means through which they could improve their quality of life without suffering.

## End of an era

Saturday 5 June 1858 was a busy working day for Snow: he gave chloroform for an operating list at King's College, several private operations and teeth extractions, and recorded each administration in his casebooks as usual. But these were to be his last entries. The following Wednesday, whilst completing *On Chloroform*, Snow suffered a stroke: he never recovered and died a week later; the post-mortem revealed signs of kidney disease and tuberculosis.[95] He was only 45 years of age and his early demise shocked and saddened the medical community. However, although Snow's death marked the end of an era – none of his contemporaries could emulate his particular skill in blending the practice and science of anaesthesia – his influence on anaesthesia had been enormous.

A few months after Snow's death, further chloroform fatalities in the London vicinity caused concern about the risks of anaesthesia to resurface: the *Times*, in particular, expressed anxiety about its dangers.[96] But Snow's success in changing medical opinion upon the safety of anaesthesia is revealed in the *Lancet's* response to the deaths. Whereas the crisis of 1854 had caused the journal to recommend restriction of anaesthesia, by October 1858 its focus had shifted to the method of administration – in each of the fatalities chloroform had been administered on a handkerchief. Using a handkerchief was 'most unjustifiable' stressed the *Lancet*, the use of an inhaler was a 'simple precaution [that] would rob chloroform of nearly all its terrors and its dangers'.[97] A few weeks later a further fatality occurred whilst Holmes Coote, surgeon at St Bartholomew's, was administering chloroform. 'The most important precaution was omitted', cried the *Lancet* – the chloroform had again been given on a handkerchief. 'With the aid of [Dr Snow's instrument] it is possible to regulate the quantity and proportion of vapour with considerable accuracy... [no] surgeon is justified in administering chloroform, or any patient in inhaling it, by a less accurate method.'[98] This suggests that although in practice many surgeons used a handkerchief, London perceptions of chloroform had been shaped irrevocably by Snow's conviction that the chief danger of chloroform anaesthesia was overdosage and by his belief that the use of apparatus could reduce the risks of the process. A small group of practitioners, including Henry Potter whom Snow had trained at St George's, remained loyal to his principles and method.[99] We shall see in the next chapter how these doctors continued Snow's mission by promoting anaesthesia as a specialist practice.

## Anaesthesia becomes the rule

By the 1860s, the practice of selective anaesthesia had diminished and pain-relief was used for most major operations. Chloroform fatalities continued to provoke medical and public concern but most surgeons had adopted Snow's view that anaesthesia offered more benefits than risks. There is no doubt that anaesthesia had brought about substantial changes to surgical practice. Prior to ether, most surgical patients had been treated with therapies or non-invasive techniques such as poultices and very few received operations. But the introduction of anaesthesia reversed this pattern. Patients were far more willing to consent to painless operations, and surgeons were encouraged to offer more complex and protracted procedures than would have been tolerated by sensible patients.[100] Excisions, for example, gradually replaced amputation as a treatment for diseased or injured bones and joints.[101] The Scottish surgeon Syme told the British Medical Association in 1865 that 'amputation below knee is seldom required, since all diseases and injuries which were formerly held to demand it may, with few exceptions, be remedied by removing the foot at the ankle'.[102] Reconstructive procedures such as the creation of new noses, plastic operations to repair disfigurements of birth or injury, and minor operations such as the removal of small tumours or repair of fistulae had risen dramatically.[103] Thus the spread of operations performed in hospitals was reconfigured.

In parallel with this shift, there was a notable increase of operations on women and children. Because anaesthesia solved the problem of control, surgeons were willing to perform longer and more complex operations on children than they could tolerate whilst sensible.[104] Children benefited from new conservative procedures such as the replacement of lithotomy – extraction of stone from the bladder, by the less invasive procedure of lithotrity – crushing the stone and leaving its fragments to dispel. In 1842 Fergusson had stated that lithotrity was not suitable for children because they could not tolerate the pain; by 1854 he was performing lithotrity on 3-year olds.[105] The widely held belief that female patients too had a greater susceptibility to pain encouraged surgeons to offer operations to more female patients after the introduction of anaesthesia, and the figures suggest that most accepted. By the 1860s, even though more major operations were performed on male rather than female patients, the proportion of female patients receiving operations had increased dramatically.[106]

Thus within 20 years or so, the Yankee dodge was no longer perceived as a radical technique with unwarranted risk; its use had transformed the practice and experience of surgery. Patients had become accustomed to the process of breathing gas, and surgeons had grown adept at operating on insensible and motionless bodies. By 1870 few surgeons were prepared to perform, and few patients willing to tolerate, surgery without anaesthesia. For all its risks, insensibility had become an integral part of the surgical experience. The next chapter will consider how the allegiance to chloroform in British practice, in tandem with continuing chloroform fatalities, combined to establish a context from which anaesthesia emerged as a new medical specialty.

# 6
# In the Name of Safety

The emergence of anaesthesia as a specialist practice during the last decades of the nineteenth century was a peculiarly English (and mainly London-driven) phenomenon. We have seen how Snow swiftly established himself as a specialist practitioner, and after his death in 1858 this pattern was continued through the 1860s and 1870s by doctors like Henry Potter, Joseph Clover and Joseph Mills who followed his principles and method. By the 1880s, a majority of the London teaching hospitals had designated posts for administrators and it was primarily these individuals – Dudley W Buxton, Frederic Hewitt and Frederick Silk, for example – who campaigned to make the study of anaesthesia a compulsory part of medical education and who formed the first professional association of anaesthetists in 1893. By the time the jubilees of the discovery of ether and chloroform were celebrated in 1896–97, anaesthesia was a recognised specialism in England, founded on a body of knowledge and practice that was distinguished from surgery. The anaesthetist was 'a man of science' who had the experience to render any patient insensible to the pain of surgery, claimed Buxton in his 1897 oration.[1] But why did specialist anaesthesia emerge, and chiefly in England?

Until the 1900s, in Scotland, most of Europe and the United States of America, anaesthetics remained under the surgeon's control and continued to be given on a cloth or sponge. Anaesthesia was a duty performed by the surgeon in advance of the operation, or delegated to a junior member of the team – a student or nurse, for example. To explain the different trajectory taken in English practice we need to return to surgical perceptions of risk. Earlier chapters have established the surgical dislike of ether and the dominance of chloroform in Britain and we saw in Chapter 5 how, even during the 1854 crisis of confidence in the London practice of anaesthesia, there was no return to ether. Surgeons chose to accommodate the risks of chloroform in return for its efficacy. Nevertheless, patients continued to die under chloroform and this persistent trickle of fatalities caused anaesthetic safety to remain a key issue of medical and public concern in England well into the 1900s. Although London surgeons had managed risk by practising selective anaesthesia during the 1850s, by the 1870s there were few who could contemplate performing any major operation without pain-relief. So surgeons were caught on the horns of a dilemma: they were keenly aware of the dangers of chloroform, yet were not prepared to sacrifice its ease for the safety of ether. It seems likely that surgeons were supportive of specialist anaesthesia because it provided a solution to the problem of risk. From the surgeon's point of view, specialist administration of anaesthesia diverted risk away from the surgical process itself. In the event of a fatality, it would be the doctor who gave the anaesthetic who was called to attend an inquest, not the surgeon. By supporting specialist anaesthesia, surgeons could continue to benefit from the advantages of chloroform, whilst side-stepping the responsibility of administration. This argument gains strength when we consider the different attitudes to anaesthetic risk held by surgeons elsewhere.

In Scotland, for example, chloroform was the main anaesthetic agent throughout the nineteenth century and beyond. But Scottish surgeons were adamant that their method avoided the risks of London practice. Drawing on the earlier claims of Simpson and Syme, they held to the view that chloroform killed through respiration rather than the heart: the clinical focus during an administration was upon the patient's respiration rather than the pulse. Their confidence appeared to be supported: because of different legal requirements in Scotland and England, few Scottish fatalities were publicised, whereas the majority of hospital anaesthetic deaths in England during this period were reported to the Registrar General.[2] For Scottish surgeons then, the use of chloroform posed no untoward risks; they perceived it as a routine task which could safely be entrusted to a student or nurse. This perception also prevailed in other parts of the world where chloroform was given on a cloth – in the southern states of America and many places in Europe, for example.

In communities where ether was the main anaesthetic – the northern states of America and a few places in Europe – there was far less fear of the risks of anaesthesia. We have seen earlier how surgeons here responded to the early chloroform fatalities by returning to ether. Certainly in Boston, this shift was driven by surgeons' desire to avoid medical malpractice suits. Writing in 1868, Mason Warren, surgeon at the Massachusetts General Hospital, confirmed that the occurrence of fatalities had established the dangers of chloroform too clearly for him to take the 'responsibility of recommending it', and on this basis he returned to ether.[3] The safety of ether meant that single-handed surgeons could give it themselves in advance of an operation, or it could be delegated to a student or nurse. Although American surgeons were not against innovation – Warren, for example, had experimented with James Arnott's method of producing anaesthesia by a mixture of ice and salt, and also with Benjamin Ward Richardson's apparatus for using ether as a local anaesthetic by spraying it directly on to a site of the body – the use of ether was not controversial and there was little development of the technique.[4] Duncum refers to only one American ether inhaler, designed by O. H. Allis in 1874.[5]

Perceptions of anaesthetic risk then were strongest in England, partly because of the predilection for chloroform, and partly because of Snow's legacy. English anaesthetic culture had been coloured by his stress on the risks of chloroform overdosage and the safety provided by inhalers. Many of the London surgical elite during the 1850s had benefited from his clinical efficacy, and in particular he had become the regular anaesthetist of London's premier surgeon, William Fergusson. Thus Snow's practice had set a precedent for extricating anaesthetic from surgical responsibility, which was broadly supported by London surgeons. The anaesthetist emerged as a new authority in the surgical hierarchy; one who was accountable for the risks of anaesthesia. But although surgeons were tacitly accepting of specialism, in practice the new working relationship between surgeon and anaesthetist brought many tensions and raised new dilemmas of ethics and etiquette. The toleration of specialism by the autocratic London surgical elite suggests the extent of their concern about the impact of the risks of anaesthesia upon their professional culture.

Thus it seems that specialist anaesthesia emerged as a consequence of these specific local influences, rather than as a result of broader shifts across the practice as a whole. It explains international differences in anaesthetic practice: for example, why Patrick Black was appointed as chloroformist to St Bartholomew's hospital in 1852, whereas the first American appointment of a physician with special responsibilities for anaesthetics was not made until 1897 in New York.[6]

In this chapter, I show how English specialists capitalised upon the opportunity presented by surgical receptivity to specialist anaesthesia. They experimented with safer alternatives to chloroform, including nitrous oxide, which became the main anaesthetic for dental work, and although technology – inhalers, mouth gags and so on – can characterise practice, specialists emulated the surgeons and stressec irreplaceable role of clinical skills in anaesthetic practice. They carved out a set of clinical responsibilities – for example a physical examination to assess patient suitability for anaesthesia – that were distinct from those performed by the surgeon. By the 1890s, many surgeons had grown to appreciate the way in which specialist practice could aid surgical work – by producing specific levels of insensibility, for example – and supported anaesthetists in their quest for formalised training.

By the late nineteenth century, anaesthesia was established as the exemplar of medicine's power to harness the products of science for humanitarian purposes. But anaesthetic fatalities cast a shadow over this image, which specialist practice was believed to have the power to remove. Specialism did not emerge uncontested. The restriction of practice to qualified doctors, for example, was strongly opposed by many dentists who were reluctant to employ a doctor to administer anaesthetics, yet knew the viability of their practice depended upon pain-relief, and the General Medical Council was slow to make anaesthesia a required subject for medical students. Nevertheless, in England uniquely, the practices and structures of specialist anaesthesia were established by 1900.

## The 1864 Chloroform Committee

We have seen in Chapters 4 and 5 how in Britain, the risks of anaesthesia dominated 1850s debates and practice. From the first, Snow had argued that the particular dangers of chloroform could be circumvented by the use of inhalers which prevented overdosage and for this reason he opposed Simpson's method of dropping chloroform on a cloth. As we saw in Chapter 3, the different methods of administration were polarised and used to characterise the differences between English and Scottish medicine. Certainly the London medical journals had been active in their promotion and support of Snow's principles during his lifetime, but the continuing incidence of 'death upon death, in cases where skilfully constructed inhalers were employed', had lessened the strength of Snow's argument.[7] The 1860s saw a significant shift in medical attitudes to the use of anaesthesia. The question of whether or not anaesthesia *should* be employed during surgery changed to *how* it could be employed more safely. The first evidence of this shift came in the report of the Chloroform Committee of the Royal Medical and Chirugical Society, published in 1864. The Committee had been set up in 1862 to examine the physiological, surgical and obstetrical effects of ether and chloroform. At this point, 123 chloroform fatalities had been identified worldwide although a significantly larger number were believed to have occurred. Drawing on a series of physiological experiments on animals, their report found – as Snow had done in 1847 – ether to be the safer agent whilst chloroform was the most efficacious. Many aspects of the report reiterated views originally voiced in the 1850s, but its conclusion that 'an apparatus is not essential to safety if due care be taken in giving the chloroform' suggested that it was not so much the method as the skill of the administrator that was the key to safe anaesthesia.[8]

Nevertheless, a core of London practitioners remained committed to the use of anaesthetic technology, including Joseph Clover, chloroformist at University College hospital who performed the experiments for the Chloroform Committee, the dentist Alfred Coleman, and Arthur Sansom who had worked with Snow at King's. Although their use of inhalers stemmed from concerns over safety, it also seems that adapting and innovating technology became a means through which specialists could promote their expertise across the wider medical community. Specialists justified their development of new apparatus through claims of increased efficiency or benefits to surgeons or patients. In 1865, for example, Sansom modified Snow's chloroform inhaler in order to remove 'certain objections' from its design such as the 'heavy and cumbrous' water bath.[9] As in the case of surgical instruments, new inhalers were named after their originator and if they achieved good reviews in the medical journals could help build a reputation. Clover designed a chloroform apparatus following Snow's suggestion that the chloroform vapour could be contained in a bag or balloon (Figure 6.1). It was highly rated by the medical journals and in 1862 was exhibited in London at the International Exhibition.[10]

The second shift marked by the 1864 Report was that rather than curtailing the use of anaesthesia through the exclusion of specific surgical procedures or particular constitutional conditions, the Committee sought to diminish the dangers of chloroform by combining it with ether.[11] It recommended combinations of ether and chloroform in two different ratios and a mixture of alcohol, chloroform and ether which became known as ACE. Trials, carried out in 70 operations across London, showed that the combined gases were safe and effective, although slower to establish insensibility.[12] The ACE was 'as good as chloroform' with less risk of post-operative vomiting, said Thomas Bryant, surgeon at Guy's and Committee member.[13] Sansom experimented with ether and developed the sequential use of chloroform followed by ether to

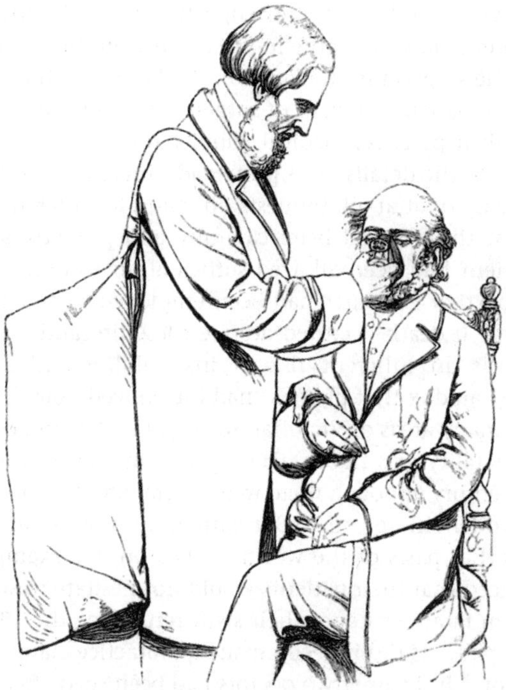

Mr. Clover's Apparatus.

*Figure 6.1* Clover demonstrates his chloroform apparatus. Taken from Sansome 1865, p. 122. Reproduced by courtesy of the Director and University Librarian, The John Rylands University Library, The University of Manchester.

maintain anaesthesia in prolonged operations.[14] Clover developed inhalers for the sequential administration of chloroform and ether as did Robert Ellis, obstetrician at the Chelsea and Belgrave Dispensary.[15] But overall, the Committee's findings made little impact on practice.

Remarkable for its absence in the report was any suggestion that chloroform should be abandoned in favour of ether. Instead, the Committee concurred with 'general opinion' that the disagreeable odour of ether, the length of time it took to establish insensibility and the excitement created by the gas made it 'inconvenient' in surgical practice.[16] This view is significant because it revealed medical unanimity that on balance, chloroform remained of net benefit to patients. It was sustainable only in a society where patients were tolerant of doctors employing practices that might have uncontrollable consequences.

Indeed, this view had been part of English anaesthetic culture from the 1840s. Chapter 2 noted how the inquest ruling on the first ether fatality exonerated the surgeon from blame on the basis that the purpose of its use – to save a patient the pain of surgery – was appropriate. From this point in English practice, medical liability for the risks of anaesthesia was confined to the details of its proper administration, rather than the decision to employ it at all. Inquests on chloroform fatalities sought to establish that the gas had been carefully and properly administered, that the patient had been allowed sufficient air to exclude the risk of asphyxia, and that measures had been employed to avoid the dangers of overdosage. If death occurred despite these precautions, then it was rationalised as an inherent risk of insensibility and as Chapter 4 showed, the paradox that progress had introduced new risks to life was common to many areas of Victorian society. Legal action concerning an anaesthetic fatality was never taken against British doctors during the nineteenth century although there were several prosecutions for violation of women under chloroform.[17] But patient attitudes to anaesthetic risk differed in other parts of the world. In Boston, for example, surgeons were convinced that their patients would not hesitate to sue in the case of chloroform fatalities hence their swift return to ether. Their instinct was correct: patients did indeed pursue malpractice claims over the use of chloroform.[18] In France, two doctors had been sued after a chloroform fatality in 1851 for 'causing death by imprudence'.[19] The doctors were eventually cleared of charges, although it seems that this case may have caused many French surgeons to develop a practice of administering chloroform for surgery to a lesser depth than Snow's recommended fourth degree, as a strategy to reduce risk of overdosage.[20]

## Dentistry and nitrous oxide

One of the most controversial aspects of fatalities was the way in which death could occur during the most 'trivial' of procedures such as the extraction of a single tooth. Dentists were acutely aware of the risks but many patients refused to consent to treatment without chloroform. If fatalities did occur during dental work they could devastate the dentist's future practice and his professional confidence. Ward Richardson recollected the way in which one of his dentist friends had become an 'old man in an hour' after a lady had died from chloroform during a tooth extraction.[21] So for dental surgeons as for general surgeons, specialist anaesthesia emerged as a method of managing the risks of insensibility. Charles Fox, surgeon at the London Dental hospital, would only work under chloroform at the patient's or the administrator's house; he believed the blame 'stuck' to the dentist if the operation had taken place at his own house. Only when Clover administered the chloroform was he agreeable to performing the work at his own premises.[22] Elite dentists who aspired to reform dentistry along similar lines to those of the medical profession were concerned that the careless use of chloroform could prejudice their claims. In 1858, John Smith, surgeon dentist at the Royal Dispensary in Edinburgh, commented that the free use of chloroform 'by the most ignorant' of dental practitioners was one of the major threats to this course.[23] The issue was discussed frequently at meetings of the Odontological Society – established by London's elite surgeon dentists in 1856 – but in 1868, some resolution was achieved by the revival of nitrous oxide.

Whilst the Chloroform Committee was pursuing its investigations, there had been a curious instance of deja vu in America. Gardner Quincy Colton, whose popular 'laughing gas' show had inspired Wells to trial nitrous oxide in 1845, reintroduced the anaesthetic to dentistry.[24] By 1863, Colton had abandoned his roadshow and established the Colton Dental Association in New York. News of this American revival first reached Britain during 1864,[25] but it was not until 1868 that nitrous oxide entered London anaesthetic practice, again through the serendipitous intervention of an American dentist, T. W. Evans, living in Paris.[26] In March 1868, Evans demonstrated nitrous oxide at the Dental Hospital of London and Moorfields Eye Hospital in front of many London surgeons and dentists. The general impression was very favourable, although Clover noted that 'a few patients were evidently insufficiently narcotised', and some displayed alarming signs of 'lividity and convulsive movements'. The 'quickness of action and the fugitive character of the insensibility' induced by the gas meant that it offered dentists 'a most extraordinary and ready mode of rendering the dental patient insensible to the agony of extraction', explained Kidd in a letter to the *British Journal of Dental Science*.[27] Within a week of seeing Evans' demonstration, Coleman, who was reluctant to use chloroform unless the patient was having more than two extractions, had developed an apparatus, produced the gas and given it successfully to four patients.[28] Nitrous oxide was a notoriously difficult gas to produce in a stable and portable form but the technological problems were quickly resolved. Within a short time, apparatus had been adapted to administer it to patients in dental surgeries and hospitals; it was adopted as widely in the provinces as in London. When Francis Kilvert visited his dentist, Mr McAdam in Hereford in 1871, he was shown the apparatus for giving the new anaesthetic 'laughing gas', which was 'much safer than chloroform, indeed quite safe'.[29]

But not all doctors were in favour of nitrous oxide. Ward Richardson, at the time President of the Medical Society of London, made a vigorous and heartfelt rebuttal of the 'childish excitement' which had greeted its revival. It was not 'an unknown, wonderful and perfectly harmless agent', he argued, rather it was 'one of the best known, least wonderful, and most dangerous of all substances that had been applied for the production of anaesthesia'.[30] His concerns stemmed from his belief that nitrous oxide was an 'asphyxiating agent', hence carried the promise of 'certain disaster'.[31] And certainly, patients inhaling nitrous oxide displayed all the signs of asphyxia; lividity of the face and convulsive movements. But London interest in nitrous oxide was sustained by Evans' donation of £100 to the Odontological Society to fund further research; a joint committee of dentists and surgeons, supported by Clover and Henry Potter, chloroformist at St George's, was set up; their preliminary report in December 1868 suggested it was highly efficient and apparently safe for short operations, but difficult to sustain in lengthier cases.[32] However, well before the final report appeared in 1872, the gas had become established as the primary anaesthetic for dentistry and also for short surgical procedures.[33]

Experienced administrators like Clover and Coleman devised new methods of giving the gas through the nose which made it useful in operations on the mouth.[34] But the main difficulty arose from the asphyxial symptoms produced by nitrous oxide which made it impossible for administrators to maintain anaesthesia for more than a few minutes. Some developed random techniques of stopping and restarting inhalation, so as to allow the symptoms to recede, and in 1878 Paul Bert, professor of physiology at the Sorbonne in Paris, showed how this problem could be overcome by adding oxygen to the gas. But this did not become widespread practice and most administrators returned to chloroform for all but dental work.[35]

Specialists rapidly took up nitrous oxide and solved the technical difficulties of producing and supplying the gas, but little attention was paid to its physiological basis.[36] It was revealing of the way in which for most doctors, anaesthesia was an empirical rather than a scientific practice. Fox, surgeon at the London Dental Hospital, spoke for many when he said that the question of whether or not nitrous oxide gas was a 'true anaesthetic' was irrelevant. Instead, he continued, the fact 'that it possesses the power of inducing a condition in which painful operations may be effected without the patient's cognisance is… undeniable… and sufficient for my present purpose'.[37]

The constant risk of chloroform fatalities spurred specialists to trial other agents; Ward Richardson, for example, had introduced bichloride of methyl into practice in 1867.[38] It was a more volatile liquid than chloroform and the surgeons at Guy's hospital used it for minor procedures such as the removal of an epithelium of the face, which only lasted one and a half minutes. On other occasions it was used sequentially with chloroform, but it did not always succeed in producing insensibility.[39] But it was not until the 1870s that chloroform's supremacy was challenged by an American surgeon who encouraged his London peers to retry ether.

## The revival of ether

Joy Jeffries, an ophthalmic surgeon from Boston, had trained in Europe in the late 1850s and had noted the general reluctance of British and European surgeons towards the use of ether. In August 1872, he attended an international congress of ophthalmologists in London and presented a paper on *Ether in ophthalmic surgery*.[40] Although Jeffries drew attention to ether's clinical advantages – less post-operative vomiting and nausea – his central argument was clear: 'I do not desire to run the risk of killing a patient with chloroform', he said.[41] He stressed the pertinence of this decision to ophthalmic surgery, where the risk associated with procedures such as cataract, iridectomy or removal of the globe was negligible. For this reason, he continued:

> I should not care to have a patient die from chloroform under my hands, and be myself tried for manslaughter afterwards. The prosecuting attorney could put scores of surgeons on the stand, whose evidence to the jury would be unanimous that I might have employed ether, which is not fatal, and hence the responsibility of the fatality of chloroform rested entirely upon me. It would be an ugly case.[42]

Thus Boston surgeons believed that the choice of British surgeons to persist in the use of chloroform was professionally untenable. Indeed, the difference revealed the way in which litigation played far less of a role in British medical culture than in North American practice, where cases of medical malpractice had risen dramatically since the 1830s.[43]

Jeffries was invited to demonstrate ether at several of the key London teaching hospitals – St George's, King's and Guy's – to see, said one surgeon, if 'our past dissatisfaction with it may be in any way due to our faults of administration'.[44] The skill of administration, said Jeffries, was to pour plenty of ether on to a napkin or sponge which was pushed into a rolled towel, and then to hold the patient down 'by main force' if necessary, until they became unconscious. Patients should be warned in advance that although they might experience sensations of choking or suffocation they should continue to take long breaths. If at any point they struggled, or asked for 'respite and fresh air', the administrator must not yield, or try to reason 'with adults excited by the anaesthetic'.[45] The *Lancet* noted that Jeffries' use of ether had a 'freedom that at first sight seemed almost startling' but it certainly produced very good results and the surgeons who had performed operations upon the etherised patients were fully satisfied with both the insensibility and the muscular relaxation it produced.[46]

Jeffries' ether demonstrations motivated many surgeons to retry ether and some, like Cooper Forster at Guy's, adopted it immediately. He had witnessed many instances of near death from chloroform when only artificial respiration and galvanism had restored life, and he was more than willing to make the shift on grounds of safety. Operation notes from Guy's show that although in some instances the patient 'took the ether very quietly', others laughed and struggled, and some suffered 'blueness, sweating and sickness'. These descriptions were far more reminiscent of the early experiences of ether, than of Jeffries' troublefree administrations. Indeed, Cooper Forster was the only surgeon at Guy's to change his practice; after experimenting with ether, Davies Colley, Durham, Howse and Poland continued to use chloroform for 100 per cent of their anaesthetic cases.[47]

Over the next few years, the use of ether became more common as the American method of giving the gas freely and with force, if necessary, spread through English practice. For many surgeons ether became a viable alternative, especially in cases where chloroform was judged to pose a particular risk – patients with heart disease for example.[48] Charles Bell Taylor, surgeon to the Nottingham and Midland Eye Infirmary, employed both chloroform and bichloride of methylene but, having learnt of the American method of giving ether from Jeffries at the Ophthalmic Congress, he adopted it as a fallback.[49]

By the late 1870s the spectrum of anaesthetic gases had broadened considerably. Specialists could now choose from combination mixtures, nitrous oxide, ether, bichloride of methyl, and, of course, chloroform. This expansion in anaesthetic gases had occurred primarily as a consequence of the push to find a safer alternative to chloroform. Nevertheless, it had recast understandings of specialist practice. Whereas in the 1850s, the use of inhalers to administer chloroform had been the mark of specialist practice, now it was the skill of the administrator to match the particular characteristics of an anaesthetic gas to the individual patient and to the specific clinical context. This shift is evident in the findings of a survey on anaesthetic practice within London hospitals carried out by the *BMJ* in 1876. The returns show the beginnings of a consensus on the appropriate use of the different gases and express the view that the single-handed administration of anaesthesia by the surgeon prior to operating was unsafe. Sansom, by now, practising at the London Hospital, gave all children under 7 years chloroform (often mixed with alcohol) dropped on to a handkerchief; nitrous oxide, using an apparatus, was given for dental procedures or very short operations; other cases were given chloroform using an inhaler, followed by ether for prolonged operations; and patients with suspected cardiac disease, alcoholism, shock or dislocations were given ether alone.[50]

In Scotland and many of the provinces, chloroform continued to be given on a cloth, although by the 1870s, some doctors had replaced the cloth with face-masks. Simpson had adjusted his method in 1860, by reducing the folded towel or cloth to 'one single layer'. He promoted this revised method with enthusiasm, extolling its ability to produce perfect anaesthesia whilst saving on the amount of chloroform used. That Simpson made this shift because of the 'careless manner in which, in particular, students and young practitioners sometimes employ the damp folded cloth over the patient's face without admitting a sufficient supply of air' suggests he had become concerned about the dangers of chloroform.[51]

Liverpool physician, Thomas Skinner, further developed Simpson's modification and designed a wire frame covered with a pad of wool and cotton which was used to cover the lower half of the patient's face. The chloroform was then dropped onto the material from a specially adapted bottle which restricted the amount that could be released by each inversion. This mask became known as Skinner's inhaler and was very popular throughout Britain and Europe.[52] As simple to use as a cloth but described as an inhaler, it suited those doctors who found the dose-based technology of London specialists hard to manage but wanted nevertheless to practice with the accoutrements of 'scientific' medicine; it chimed with the Victorian appreciation of 'gadgetry'.[53]

We see how the Scottish method continued to follow Simpson's principles through the writings of Syme's son-in-law, Joseph Lister, who became professor of surgery at Glasgow in 1860, and wrote articles on anaesthesia for three successive editions of Holmes' *System of Surgery* published in 1861, 1870 and 1882. 'The very prevalent opinion that the pulse is the most important symptom in the administration of chloroform is certainly a most serious mistake,' wrote Lister in 1861. 'As a general rule, the safety of the patient will be most promoted by disregarding it altogether', he continued, 'so that the attention may be devoted exclusively to the breathing.'[54] In 1870, Lister claimed that during the previous 9 years there had been no chloroform fatalities at either Edinburgh or Glasgow Infirmary: 'yet…a folded towel…is still the only apparatus employed…preliminary examination of the heart is never thought of…the pulse is entirely disregarded; but vigilant attention is kept upon the respiration'.[55] But by the time Lister contributed his third article in 1882 he had moved from Edinburgh to become Professor of Surgery at King's College and had re-worked the details of the Scottish method of administration. Having experimented with Skinner's mask but found it was 'liable to the danger of giving the chloroform too strong', he described making trials using a piece of flannel stretched over a frame. His aim was to produce a 'simple manner' of administration that could be adopted by practitioners throughout the country, many of whom feared the complications of special apparatus.[56] Here, perhaps, was an Anglo-Scottish compromise. It suggests that Lister had been influenced both by a chloroform fatality which had occurred in one of his private patients, and by the climate of London anaesthetic practice with its stress on the dangers of chloroform overdosage and its use of a panoply of technology.[57] We turn now to see how during this period of experimentation with new gases, the emerging specialists drew together a set of specific technical skills and clinical responsibilities that came to define the role of an anaesthetist.

## The making of the anaesthetist

In 1852, Snow sketched out the rationale for specialist practice:

> However trivial a matter the exhibition of chloroform may have been considered by many on its first introduction, I believe there is no one who does not now look on it as a subject requiring the utmost care and attention, together with a thorough knowledge of all the symptoms it may induce.…While fully acknowledging the great attainments of the gentlemen who usually occupy the office of dresser or house-surgeon in the London hospitals, I consider that there are grave objections to their having charge of the chloroform.…No person ought to administer chloroform without first making its action a subject of special attention; and, as there

requires to be some one always on the spot to administer it on emergency at the public hospitals, it should be the duty of a permanent resident medical officer.[58]

Technology became the means through which specialists extended control over the anaesthetic process. Nasal intubation, for instance, was developed to improve the depth of insensibility during operations on the mouth. In 1878, St Bartholomew's anaesthetist, Joseph Mills described giving a chloroform mixture through the rubber tube of a Junker's inhaler which was passed through the nostril and round into the pharynx. He suggested that it was also possible to administer gas through a flexible catheter held in the mouth.[59] At the International Congress of Medicine held in London in 1881, Clover exhibited new instruments for maintaining anaesthesia during operations in the mouth: 'a bag with a nasal tube adjusted and firmly fixed into the nostril with a screw movement; also funnel-shaped india rubber tubes for conveying the anaesthetic to the back of the mouth during operations on the jaw'.[60] Snow had described his use of a mouth-prop which was inserted between the teeth and used to prevent muscle rigours closing the mouth whilst the patient was under the influence of chloroform, but new gags were introduced and Thomas Smith designed one specifically for use in cleft palate operations.[61]

Specialists adopted established surgical techniques for restoring the airway – tracheotomy and laryngotomy, for example – which introduced a tube into the airway by means of an incision in the neck. In 1877, Clover described inserting a small, curved cannula into the crico-thyroid membrane to restore life to a patient suffering from total respiratory malfunction.[62] Snow had used such means to successfully resuscitate animals from an overdose of chloroform when artificial respiration failed,[63] and there are examples of tracheotomies being performed as a last resort in chloroform fatalities from the 1850s, but at this point it was the surgeon, rather than the administrator who undertook the procedure. Over time, this development led to new initiatives such as using the opening to suck out debris and blood in the case of haemorrhage during operations on the mouth or nose, or administering oxygen gas to aid artificial respiration – techniques again performed by the specialist rather than the surgeon.[64]

Henry R. Silvester's two-part method of resuscitating chloroform casualties became one of the most popular techniques. The first movement to 'immediately and suddenly compress the front and sides of the chest by the patient's own arms' sought to expel 'the poisoning vapour'.

Then the arms were raised and stretched in order to 'imitate natural respiration' (Figure 6.2a and b). Silvester advised that the movements should be performed 'with long perseverance and unfailing regularity'.

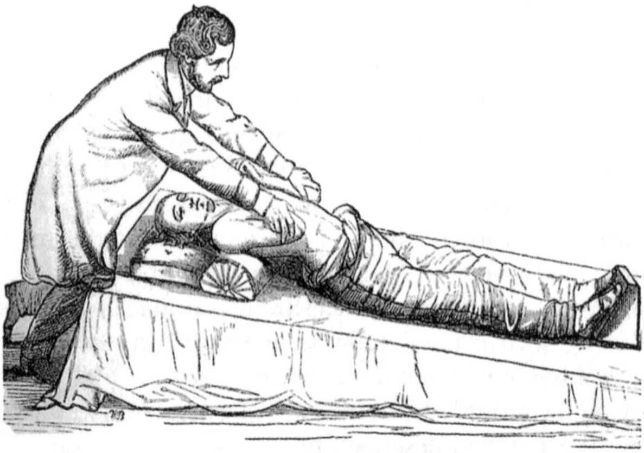

**Dr. Silvester's method of resuscitation. First effort, to expel air from the chest.**

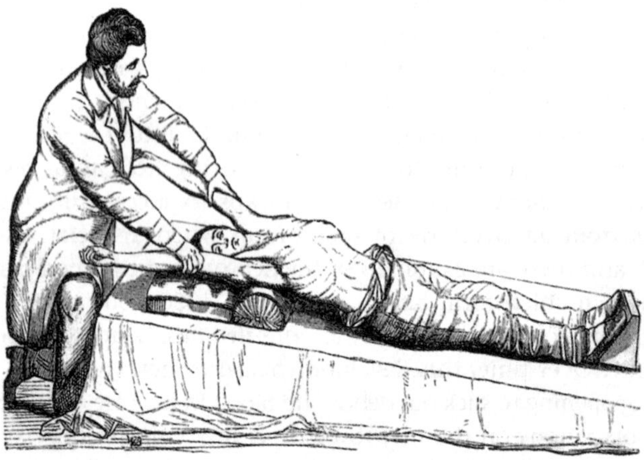

**Dr. Silvester's method. Producing expansion of the chest.**

*Figure 6.2a and b* Henry R. Silvester's two-part method of resuscitating chloroform casualities. *MT* 1 (1863) 389. Taken from Sansome 1865, pp. 109–10. Reproduced by Courtesy of The Director and University Librarian, The John Rylands University Library. The University of Manchester.

Pre- and post-operative pain-relief also became redefined as an anaesthetic responsibility. Prior to ether, surgeons often gave patients opiates to ease the pain of the wound and during the 1850s, Snow described the occasional use of morphine by a surgeon before an operation in an attempt to diminish the post-operative pain. In the late 1850s attention had turned to the possibility of using narcotics for local pain-relief by administering them subcutaneously: Edinburgh physician, Alexander Wood and Charles Hunter of St George's Hospital described injecting morphine using a glass syringe and hollow needle, and the technique was quickly established as a treatment for neuralgia.[65] French physiologist Claude Bernard's experiments of the 1860s showed that the combined use of morphine and chloroform – 'mixed anaesthesia' – diminished excitement and reduced post-operative vomiting and pain.[66] Pre-operative administration of subcutaneous morphia became common practice amongst specialists,[67] and this technique was extended during the 1890s by the addition of atropine – a cardiac stimulant.[68] Although the purpose of these injections was to diminish vomiting and pain, specialists noted the effective way in which the combination calmed frightened patients.[69] By 1900 the administration of pre-operative injections of scopolamine (a derivative of atropine) and morphine was common amongst London specialists; Dudley Buxton recommended them for 'terrified patients'.[70]

One of the best ways to appreciate the growth in the range and complexity of anaesthetic methods and technology during this period is by considering the emergence of the anaesthetist's table. At the time of Snow's death in 1858, his equipment for giving chloroform consisted of an inhaler – large or portable version – a bottle of chloroform and a mouth-stop; inhalers were listed in surgical catalogues from the late 1840s.[71] But, by the 1890s, a designated table, equipped with anaesthetic liquids and equipment was becoming part of standard operating theatre equipment in London and the larger provincial hospitals. It was a visible and tangible representation of the way in which the Yankee dodge had become established as a new medical function. An 1899 volume on surgical ward work and nursing described the contents of an average anaesthetist's table. It was said to contain two bottles each of chloroform and ether, alcohol for mixing with ether, a dropper bottle, a folded towel, a face-mask, a Clover or Junkers' ether inhaler, a small bottle of Vaseline for protection of the patient's face from volatile liquids, tongue forceps, clean towel, small basin, hypodermic syringe filled with ether that would be injected subcutaneously in the event of the heart's action becoming feeble, other syringes to give pre-medication injections of morphine and atropine, eucalyptus oil for sterilizing the syringes, a strong solution of ammonia as a stimulant, capsules of amyl nitrate which again would be used to stimulate the heart or respiration and cocaine solutions for local anaesthesia by injection.[72] The contrast speaks for itself.

However, although technology grew to characterise specialism, anaesthetists did not found their claims for status and authority upon this aspect of their practice. Instead, they focused on the way in which successful anaesthesia was highly dependent upon the administrator's skill in managing patients' apprehension of, and response to gas inhalation. A calm patient became insensible quickly and quietly; a fearful patient with erratic breathing was likely to struggle against the sensations of the gas. In this way, specialists elevated the status of the administration of anaesthesia from a manual task – even one based on technological expertise – to a process requiring the same characteristics of judgement, skill and patient empathy as those held by physicians and surgeons. And nowhere were these skills more highly in demand than in managing patients' fear of anaesthesia.

We saw in Chapter 4 how patients' fear of surgical intervention in pre-anaesthetic times had metamorphosed into a particular fear of chloroform during the 1850s and was acknowledged within the surgeon's calculus of risk. Although patients chose to breathe gas for every type of procedure, from the nominally 'trivial' removal of toe nails or cysts, to the more complex procedures of amputations or breast removals, chloroform fatalities continued to command wide publicity during the second half of the nineteenth century and patients remained fearful of sudden death under anaesthesia. Anaesthetists claimed that the 'bedside' manner of a specialist – an ability to instill confidence and cooperation in the patient – was vital to the success of each inhalation. It was imperative to speak 'kindly' to patients, said G. C. Coles, chloroformist at the Great Northern Hospital.[73] Holland, anaesthetic administrator at the Hospital for Women, explained how he visited the patient before an operation and with 'a cheerful and confident bearing', listened to the heart, percussed the lungs and made a 'judgement' on the patient's peculiarities.[74] He was explicit that it was of psychological rather than clinical benefit; it reassured patients that the outlook was good and served as a strong public representation of anaesthetic skill. When a young man died under chloroform at the Middlesex hospital in 1891, his friends made a particular complaint to the coroner at the subsequent inquest that 'the administrator did not take the precaution to examine the condition of the heart', although this did not change the eventual verdict that exonerated the administrator from blame, and ruled that death had occurred because of the effects of chloroform.[75]

As well as calming pre-operative anxieties, specialists also stage-managed the environment in which the inhalation took place so as to avoid frightening patients. Although apparatus became a mark of specialist practice it paradoxically appeared to exacerbate patient fear of anaesthesia; Simpson had claimed from the late 1840s that patients were frightened of inhalers. Specialists were keenly aware of this tension and sought to veil the presence of anaesthetic technology, and create a space that on entering appeared 'normal' to patients. The best dental premises, commented Arthur Underwood, administrator at the London Dental hospital, were those presented like ordinary sitting rooms, with an armchair for the patient to sit in and 'no instruments visible'.[76] It was essential, said Frederic Hewitt, anaesthetist at the London hospital, that the patient was not 'alarmed by the sight or the arrangement of the somewhat complicated apparatus'.[77] Hewitt described how when administering sequential nitrous oxide and ether, he filled the gas-bag with nitrous oxide, and hid it under the operating table or bed. The patient was admitted into the room and lay down 'without having seen any of the apparatus'. Hewitt positioned the face-piece and began to give air, unmixed with any ether. When the patient's confidence was gained, he then 'noiselessly adapted' the gas-bag to give nitrous oxide, and gradually replaced this with ether.[78] One of the problems created by the use of nitrous oxide was the difficulty of disguising large cylinders of gas. The *British Journal of Dental Science* advised dentists to partition off their rooms to conceal them. In this way, the patient would see nothing more 'alarming' than 'a few feet of India-rubber tubing, with Mr Clover's face-piece attached'.[79] Thus, an 'apparent absence of precautions, paraphernalia, anxious enquiring glances, and mysterious hints between the operator and the anaesthetist' served to reassure the patient 'that there is really nothing to make a fuss about', affirmed Underwood.[80]

In hospital practice it was more difficult for anaesthetists to control the environment, particularly when anaesthesia was given in the operating theatre. Snow had noted the benefits of anaesthetising patients behind a screen. It was a way of using anaesthesia not just to protect the physical sensibilities of vulnerable female patients, he suggested, but also their moral feelings. He continued:

> There are many operations on the female which medical students could seldom witness except at the expense of some shock to the feelings of the patient. They are now generally conducted in the

> hospitals in this wise; the patient inhales and becomes insensible whilst only one or two surgeons and the nurse are present in the private ward, or behind the screen with her; the students then come in and witness the operation, and go away again before the consciousness of the patient has returned.[81]

He suggested that the same method could be used for females undergoing gynaecological operations; their legs would then not be opened and strapped to the table until they were protected by unconsciousness. By the end of the nineteenth century a few London hospitals had begun to set aside anterooms so that patients could be lulled into unconsciousness away from the hustle and bustle of the theatre.[82] Such arrangements stemmed from practical concerns – it was agreed that patients responded more quickly to anaesthesia in a quiet environment, for instance – but resulted in a clear demarcation between anaesthetic and surgical space, further emphasised by the emergence of the anaesthetist's table which gave tangible evidence of the new function (Figure 6.3).

From the anaesthetist's point of view one of the most impressive ways to demonstrate expertise was to minimise any excitement or

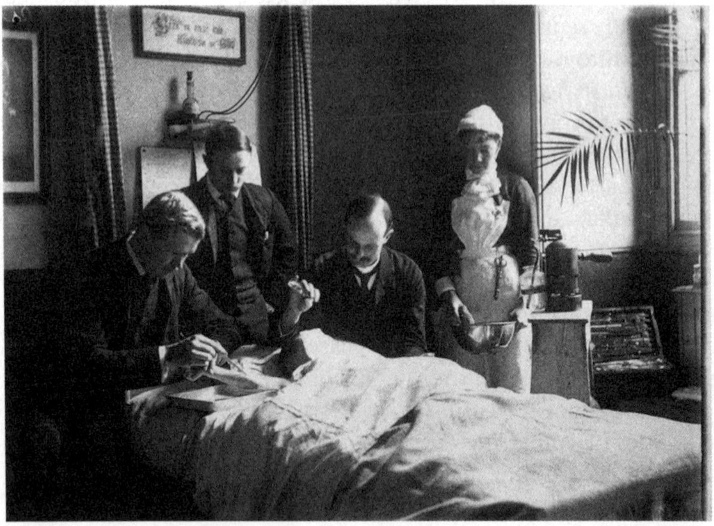

*Figure 6.3* Anaesthesia being given to a patient in a ward bed at St Bartholomew's Hospital c.1890. Reproduced by kind permission of St Bartholomew's Hospital Archives and Museum.

struggling during the inhalation so that patients slipped into unconsciousness quickly and quietly. Textbooks frequently referred to the 'excitement' stage of inhaling gases which revealed the way in which the struggling or hysteria of patients inhaling ether or chloroform, which had threatened the social norms of the 1840s, had become reconstructed as the medical phenomena of anaesthesia, rather than as a social problem. By the late nineteenth century surgical resistance to controlling patients through physical restraint had diminished. We saw earlier the willingness of London surgeons to trial Jeffries' 'heroic' method of administering ether during the 1870s.[83] Anaesthetists accepted that there were occasions when force was necessary to ensure a successful inhalation. If a patient was reluctant to cooperate with the administrator during an inhalation then they must be forcibly restrained and the anaesthetic 'pushed', said Samuel Osborn, chloroformist at St Thomas' in 1881. Nevertheless, his comment that this was a 'more difficult' thing to do in private administrations suggests that particularly outside hospital practice, old views lingered.[84] Indeed, specialists usually preceded ether with either nitrous oxide or chloroform precisely to secure a smoothly engineered induction.

In the wider hospital structures the new practice of anaesthesia was developing a profile. At Guy's hospital, chloroform and ether were listed as separate items in the apothecary's shop from 1857 onwards, and between 1857 and 1864, the spend on these drugs doubled.[85] At St Bartholomew's, the amounts of chloroform recorded in hospital ledgers increased from 55lb in 1855, to an average of 241lbs between 1866 and 1875 and then to 339lbs by 1886.[86] And by the 1900s, anaesthetic records, separate to those of surgery began to be kept by some hospitals.[87] The anaesthetist and the trappings of his practice were becoming part of the fabric of medicine. But given the dynamics of elite London medicine, the emergence of specialism was only possible with the support of surgeons.

## Surgeons and anaesthetists

Although selective anaesthesia had been accepted as a strategy for reducing risk in the 1850s, by the 1870s it was almost impossible for any surgeon – particularly those aspiring to join the London medical elite – to justify the non-use of anaesthesia in major operations. At Guy's between 1872 and 1874, 98 per cent of patients undergoing surgery were offered anaesthesia and less than 1 per cent refused it.[88] Indeed it could be said that many of the developments in surgery that had come about since 1846 had emerged only because of the humanitarian and practical benefits of anaesthesia. The success of antisepsis techniques introduced from the 1870s onwards, for instance, had encouraged surgeons to operate on new areas of the body, particularly the chest and abdomen. But antisepsis was heavily dependent upon the insensibility of anaesthesia. It is hard to imagine how conscious patients could have tolerated the lengthy preliminaries of the carbolic spray whilst waiting for the first cut of the knife. Yet just as patients remained fearful of the process of unconsciousness, surgeons continued to be haunted by the risks of anaesthesia.

The powers of anaesthesia were 'fraught with subtle and imminent danger', noted Fergusson.[89] His contemporary, John Erichsen, surgeon at University College, believed the use of chloroform in private practice 'adds at least a hundred per cent. to the anxieties and responsibilities of the surgeon'.[90] Baden, ophthalmic surgeon at Guy's, who had administered 3,483 anaesthetics between 1862–69, was proud of his unblemished record yet could not help noting the 'feeling of unsafety' which accompanied every anaesthetic administration.[91] And Underwood told of several incidents at the Dental hospital where the anaesthetic had to be given twice because the 'nervousness of the surgeon induced him to repeatedly implore the anaesthetist not to give too much, thereby exciting the patient with groundless fears and flurrying the administrator'.[92]

What surgeons feared most was that their surgical reputations would be tarnished by anaesthetic fatalities. As we have seen through Snow's practice, many London surgeons were willing to use a second doctor to give anaesthesia from the late 1840s onwards. But in order to achieve a complete separation between anaesthetic and surgical risk, there had to be a shift in accountability. This was the key difference between surgeons who employed a specialist to anaesthetise patients – as in English practice – and surgeons who delegated the administration to a junior doctor or nurse – as in Scottish, American and European practice. In the first case, the specialist took on responsibility for the risks of anaesthesia, in the second the surgeon was held accountable for both anaesthetic and surgical risk. 'The person who undertakes control of the anaesthetic is responsible for the safety of the patient', affirmed the *Lancet* in 1896, although it acknowledged that some doctors maintained that 'the burden of the responsibility' was borne by the surgeon who operated.[93] Indeed, for surgeons who were less fearful of the risks of anaesthesia, the prospect of abnegating responsibility in this manner was strongly detrimental to professional values. 'Is it that a surgeon is no longer responsible for the safety of the patient, but that the responsibility is to be shared by the man who gives chloroform or ether?' exclaimed Edinburgh-trained surgeon Edward Lawrie at the turn of the century. He continued:

> In certain eventualities is the anaesthetist to dictate to the surgeon so that the surgeon becomes a mere operator, a subordinate instead of a chief, who under all circumstances retains his supreme command and the entire responsibility in his own hands? This constitutes in my opinion, the tendency to the degradation of surgery against which all surgeons should guard with all their might.[94]

The divide between Scottish and English surgeons on the value of specialism affirms the way in which it was a contingent process that occurred because of the particularities of English medicine. However, although specialism may well have emerged in this way, over time surgeons found it supported and eased their work on a more practical level. Abdominal operations, for example, were much easier to perform under 'full' anaesthesia that stopped all reflex movements; surgeons operating on the nose and throat required the laryngeal reflex to be kept functioning so patients would not choke on blood and mucus.[95] Thus surgical success came partly to depend upon anaesthetic skill.

But attempts by anaesthetists to establish equal authority led in many instances to a fraught and controversial relationship with the surgeon. Snow recorded occasions when Fergusson's impatience to commence surgery had forced him to use a handkerchief, rather than inhaler, so as to anaesthetise the patient more quickly.[96] Underwood, who gave anaesthetics at the London Dental School during the 1880s, described the limits of the surgeon's responsibility: he should check that he had room to operate, that the mouth-gag was inserted correctly and that the patient was in a suitable position for surgery. Apart from this, he simply had to 'keep quiet and be ready for the administrator's signal'.[97] Sansom explained how responsibility for patient safety was shared equally:

> between the surgeon who counsels the administration for the performance of the operation and the administrator who deals with the agent the use of which demands care and caution.[98]

But many surgeons continued to view anaesthesia as a manual task and believed their surgical requirements – in relation to the position of the patient, for example – were paramount as a royal example will show.

In June 1902, Hewitt, anaesthetist at the London and St George's hospitals, and founder member of the Society of Anaesthetists, gave anaesthesia to Edward VII whilst the surgeon, Frederick Treves, drained an abscess in the appendix. Hewitt later explained how Treves had alerted him to the possible operation and 'ordered' him to wait at home until he was called. When Treves arrived, he told Hewitt he was required 'at once', asked him to collect the nurse, and then join him at his home. When Hewitt arrived, Treves was eating a large lunch; Hewitt was not invited to share it. Treves then left for the Palace in his own carriage with Hewitt and the nurse following. (During the journey, Hewitt stopped to buy some chocolate in lieu of lunch.) Hewitt's primary concern at this time was that he should be allowed to examine the King before giving the anaesthetic. He had mentioned this to Treves, who responded by saying he could tell him all he needed to know, and 'rather hinted that any examination would be unnecessary'. On arrival the Buckingham Palace, Hewitt was relieved to see an operating table, brought in specially for the event, in the King's dressing room. Treves and the other doctors – including Thomas Smith, surgeon at St Bartholomews' and Great Ormond Street hospitals; and Thomas Barlow, professor of clinical medicine at University College hospital – had decided that the King should first be anaesthetised on his bed and then moved on a sheet to the operating table. From an anaesthetic perspective this could cause breathing difficulties in the patient during the move. Hewitt had anticipated this suggestion and decided in advance to strongly oppose such a plan. Treves agreed to his request but Hewitt had to persist in requesting a physical examination – though listening to the King's heart through a 'well-covered chest wall and bed attire' meant that he could hear no sounds. Once the administration began the Queen exclaimed about the dangers of the anaesthetic getting into the King's eye; she was persuaded to leave, though Hewitt noted that he wished he had suggested this earlier: 'the rigid stage and slight change of colour (inevitable in a patient) must have alarmed her I fear'. Smith, who assisted Treves during the operation, confided to Hewitt that he would not 'have my job for all the world', and offered to hold the King down if Hewitt did not want to take the risk of keeping him under deep anaesthesia. After a slight delay in the recovery of the King, the operation concluded successfully and the medical entourage was treated to lunch with champagne.[99]

Treves supported specialist anaesthesia and acknowledged its value, yet in practice he paid little attention to supporting Hewitt's requirements. For this reason, the nascent partnership between anaesthetist and surgeon was frequently fraught and contested, and tension persisted well into the twentieth century. The difficulty in achieving equal authority was also reinforced by the economic structures of the relationship. Throughout the nineteenth century, and for much of the twentieth, the anaesthetist was paid by the surgeon who collected his fee from the patient, and in hospital practice, salaries for anaesthetists remained lower than those offered to surgeons and physicians.[100] We are led therefore to question why men like Hewitt chose to practice in such a challenging field?

One answer, well exemplified by Snow, is that in anaesthesia, as in other new specialisms, bright young men without social connections might create a living which was richer and perhaps more satisfying than general practice, even if their earnings fell far short of elite surgeons and physicians. By time of his death in 1858, Snow practised anaesthesia full time with earnings in the region of £1000 per annum.[101] This was certainly not comparable to the wealth of elite surgeons like Astley Cooper, whose annual income ranged between £15,000 and £21,000, or Benjamin Brodie, who earned around £10,000 a year in the 1840s, or indeed William Fergusson, who earned around £8000 in 1858.[102] Nevertheless, compared to the few hundred pounds a year Snow had struggled to secure prior to anaesthesia, this income kept him in relative comfort.

Other doctors turned to anaesthesia when their original career aspirations failed. Clover, for example, aspired to be a surgeon but was prevented by his poor health. When he died in 1882 he left an estate valued at around £27,000, similar to that of many London surgeons of the period.[103] Hewitt intended to practice as a physician but poor eyesight caused him to take up anaesthetics.[104] Of course, as in other emerging specialities of the period, full-time practice of anaesthesia could only be sustained in a buoyant private practice market and for this reason there were few anaesthetists outside the metropolis who could survive without other forms of income.

Yet despite surgical support for specialism and role-models like Hewitt, many anaesthetics continued to be given by students or junior doctors who had little experience of the complexities of the process, and administered chloroform on a cloth. For surgeons familiar with the advantages of specialism, the dangers of unskilled practice were alarming. Chloroform was given 'without undoing the corset in a woman, and without examining the mouth for foreign bodies of "false teeth", without the essential equipment of anaesthesia: a gag to stop the spasmodic closure of the jaws, forceps to pull out the tongue and a tracheotomy tube', warned Marmaduke Shield, surgeon at St George's. We 'all know that by a mixture of good fortune and careful supervision on the part of the operator such cases commonly do well', he continued, but often patients are rescued 'from the jaws of death by prolonged artificial respiration'.[105] Nor did all administrators take into account that for patients, wrote Frederick Treves, surgeon at the London hospital, in 1897, the mask of the anaesthetist had become a 'symbol of the Valley of the Shadow of Death'. Treves spoke of administrators who displayed 'copious apparatus' to the patient which they manipulated 'with the stolid ostentation of an executioner', and of others who called 'jauntily for a folded handkerchief, and, after placing it over the patient's trembling face' chattered 'incontinently' of their summer holidays.[106] Because of these problems specialists turned their attention to matters of training all doctors in the basics of anaesthesia.

## Specialists as teachers

The need for anaesthetic training for medical students was raised by Walter Rigden, late medical officer to University College, in his *BMJ* anaesthetic survey return of 1876.[107] An enquiry undertaken in 1883 by the Glasgow Royal Infirmary confirmed there was little formal instruction of the subject.[108] Students learnt the practice by watching others in the operating theatre; in teaching hospitals the subject was usually included in surgical lectures. Although larger provincial hospitals had appointed anaesthetists by the 1890s, smaller hospitals and rural areas were reliant on the provision of anaesthetic services by general practitioners. One rural surgeon writing in 1900 explained that most of the cases in which a general practitioner was called to give an anaesthetic were emergencies which required such swift action that there was often no time for an examination. For all doctors, he continued, providing pain-relief for surgery was a 'bounden duty', but the 'minutiae of detail' caused most general practitioners to 'shrink' from giving an anaesthetic.[109]

It seems that the rhetoric of specialism and the complexity of technology and methods had served to intensify, rather than diminish most doctors' apprehension about anaesthetic administration, hence the consistent popularity of chloroform. Despite being the riskiest of all anaesthetics it was the easiest to give; few administrators outside the large hospitals were skilled at either combining it with ether, or using a nitrous oxide–ether sequence. Indeed Lister had argued in 1870 that specialist appointments reduced, rather than increased the safety of anaesthetic practice because they invested the administration of chloroform 'with an air of needless mystery' and served to withhold from students 'the opportunity of being trained in an important duty, which any one of them may be called upon to discharge on commencing practice'.[110] The medical profession was in accord that giving a safe and effective anaesthetic was a primary duty of all qualified medical practitioners. But the detail of how this was to be achieved occupied specialists during the 1890s.

From the 1880s onwards London specialists had expressed their concern that issues of quality and safety would undermine claims for specialism. Buxton, anaesthetist at University College hospital, published a new book on anaesthetic practice in 1888, followed in 1893 by Hewitt, who held appointments at Charing Cross, the London and the Royal Dental hospitals. Both stressed that books alone could teach only the 'rudiments' of practice. In 1893, Frederick Silk, anaesthetist at Guy's, the Royal Free hospital and the London Dental School, used a meeting of the Thames Valley branch of the British Medical Association to call for more 'systematic teaching' on the subject of anaesthetics. In London, 11 out of the 12 medical schools in the metropolis employed anaesthetists who taught the subject separately to surgery. As instruction in anaesthesia was not compulsory, Silk estimated that only around 35 per cent of London students attended this teaching. When these statistics were extended across the whole country, he guessed that fewer than 18 per cent of medical students received any instruction on anaesthetics before qualification. He proposed that anaesthetics should be accorded the same importance as other medical subjects; certification of a specific set of lectures and demonstrations, and a period of clinical practice.[111] Thus specialists sought to establish the practice of anaesthesia on a two-tier system. Specialists – those holding appointments in teaching hospitals and medical schools – should take responsibility for teaching the rudiments of practice to students and junior doctors. This would guarantee that on qualifying every doctor was capable of giving an anaesthetic. Specialists in large hospitals would be responsible for the most complex anaesthetic cases and they would also become an authority on the subject for those outside medicine, coroners, for example, who had to navigate the intricacies of anaesthetic death through inquests.[112] And in part, it was as a means of communicating these aims and goals that the first professional association of anaesthetists was set up in 1893.

Forty members joined the Society of Anaesthetists; nine from the provinces and the remainder from London. Criteria for membership were broad: an anaesthetic appointment, an anaesthetic private practice or a specialist interest. Indeed, there were several surgeons amongst the first cohort. The format of meetings followed that of other medical societies: discussions and the reading of short papers. The emergence of such a group can be read as part of the overall professionalisation process of medicine;[113] it can also evidence the pressure of an overcrowded medical market where status had a significant effect upon the ability to make a medical living, and specialism could provide a useful differentiation.[114] But neither of these explanations extends our knowledge of the specific dynamics of the specialisation of anaesthesia. To understand why Silk and his fellow anaesthetists felt a need to bond and share clinical experiences through the formation of a society, we need to turn to the broader standing of anaesthesia.

## Art or science?

From 1846, the power of medicine to remove the pain of surgery through anaesthesia had provided indisputable evidence of the way in which medicine offered the means to radically improve and enhance the quality of life. Anaesthesia was the 'grandest and most blessed of discoveries', said Charles Darwin.[115] By the 1890s, it had become one of the century's great celebrations, seen as equal to Jenner's work on vaccination and Lister's introduction of antiseptic surgery, both of which had developed as universal rather than specialist techniques. The problems encountered in effecting successful anaesthesia with ether in the 1840s had long subsided, as too had the debates on the propriety of insensibility and the need for selectivity in its use; by the 1890s, whatever the condition of the patient, if he was fit for an operation then he was considered fit for anaesthesia. Yet against this backdrop of progress and achievement, remained the disturbing paradox that anaesthesia still carried a significant risk of death and there was no consensus on the cause of fatalities. It is possible that this innate risk may well have been accommodated more easily, but for the contrast with the wider medical context.

In many areas of medicine, particularly physiology, bacteriology and surgery, the last decades of the nineteenth century were branded by optimism and progress. Surgeons, on the basis of antiseptics and anaesthesia, had been empowered as the 'frontiersmen and explorers' of new parts of the body; the abdomen, brain and chest were emerging as new sites for intervention.[116] Worboys has shown how the germ theories and practices of the period were the 'carriers of *new meanings* for science in medicine'; they created the assumption that prevention and cure of disease was possible through the knowledge of its mechanisms.[117] Such knowledge was indeed the purpose of the work of the new breed of experimental physiologists.[118] But for the anaesthetist, the absolute determinism espoused by Claude Bernard's 'experimental method' remained elusive.[119]

The difficulty of integrating the findings of experimental physiology to the practice of anaesthesia had been long recognised. Responding to the publication of the 1864 Chloroform Committee Report, Charles Kidd highlighted the gulf that existed between experimental work on lower animals and the clinical treatment of patients. Deaths in the human subject, he said, were due to idosyncrasies which did not exist in the lower animals, and for this reason, physiological experiments 'such as dropping chloroform on the heart of a frog, or throwing it into the peritoneum, are fallacious; the experiment kills, but not the chloroform'. Nor could physiology prevent errors of administration, or control the effects of the patient's emotions or fear.[120] Ward Richardson found his method of artificial respiration perfectly revived 'dead' animals, yet was far less predictable on patients.[121] In the clinical context, said Buxton,

> it is practically impossible to say who has a healthy heart or . . . who has a resistive heart. Abuse of living, hard living, scanty diet, terror, wasting disease, are factors in the problem which face the anaesthetist . . . the physiologist cares for none of these things.[122]

The difficulties were recounted by Hare and Thornton in their study of the influence of chloroform upon the respiration and circulation, which had formed part of the Hyderabad Commission of 1890:

> The variation in the action of a drug on a diseased individual from its effect on the normal one is notorious, and we have no right to dogmatically assert that there is no danger of circulatory depression in man even if we found no evidence of failure in dogs, because there may be many idiosyncrasies or variations, through disease in the human being which may completely reverse the results of experiments on healthy animals.[123]

Experimental physiology had failed to provide the 'magic bullet' that would remove risk from anaesthesia. The reductionist approaches and mechanistic solutions which seemed to have played such a vital part in furthering the development of surgery and bacteriology had proved to have little value in the case of anaesthesia. Administering anaesthesia involved the softer – but superior – arts of clinical judgement and skill.

The paradoxical status of anaesthesia – was it an art or a science? – is evident in the many retrospectives which appeared during the jubilees of ether and chloroform in 1896–97. Dudley Buxton, president of the Society of Anaesthetists, acknowledged its ambiguous status in his jubilee oration. How could the apparently innate risk of anaesthesia be integrated in a medical culture in which certainty and predictability had become the new watchwords? Buxton was adamant that the factor that made the difference to anaesthetic risk was the anaesthetist: an 'accomplished chloroformist will use the rudest means of bringing about chloroform sleep with impunity', he explained. The anaesthetist must become a 'man of science' through developing a body of knowledge of the physiology of the action of anaesthetic agents upon the human body that gave him the power to:

> avoid recognised dangers, to guard against their coming and be ready to counteract them when they arise and to guide his patient through safe paths to anaesthesia.[124]

But this was not science in its laboratory meaning, rather it grew out of the 'most important criterion' of late Victorian medicine – clinical experience.[125] The anaesthetist was required to study each case and adapt the practice of his specialism to 'the ever-varying conditions of that ever-changing problem – the human body'.[126]

This was the purpose of the Society of Anaesthetists. It was a 'laboratory' in which the 'science' of clinical practice could be shared; the majority of papers and debates focused on the practice and technology of anaesthesia.[127] By the 1890s, the minutiae of anaesthetic practice was outside the experience of most doctors. And although a surgeon in London and in the larger provincial hospitals was assisted by a team, the anaesthetist was a lone figure. Sharing the problems of successfully anaesthetising, say, 'the healthy looking schoolboy with enlarged tonsils and post-nasal growths' or 'the drayman with angina Ludovici' with other anaesthetists, helped sustain specialists in their daily dice with the risk of anaesthetic death and new expectations of efficacy.[128]

Hewitt was appointed Emeritus lecturer of anaesthetics at the London hospital in 1903 – the same year that the General Medical Council added a requirement for anaesthetic training to its regulations. His first lecture distanced current practice from that of the 1840s and 1850s, when anaesthetic failures or 'any recalcitrant behaviour, unconscious struggling, movement, rigidity, embarrassed breathing, pulse failure, or undesirable after-effects were attributed to the patient'. Now, he said, such matters had shifted to the province of the anaesthetist: 'whatever the special circumstances of a case may be a smooth and satisfactory anaesthetic may, with the rarest of exceptions, be depended upon'.[129] Specialists were in agreement that it was possible to classify patients into different anaesthetic types: good, normal, bad or difficult subjects, and to use these classifications as a basis for choice of gas and method. But the skills that were needed to make such categorisations derived from clinical experience, not from textbooks or rote learning. For instance, the types of patients that might be problematic from a surgical or medical point of view were often the best to anaesthetise. The 'perfect type' of patient for anaesthesia, noted Hewitt, was a middle-aged woman in moderately good health, of spare build, sallow complexion, placid temperament and moderate habits, possessing a free nasal airway and somewhat defective teeth. The 'John Bull' type, however, a middle-aged man of powerful build, with red shiny cheeks, thick necks and good teeth – a more 'healthy specimen' – required far more careful handling. By the 1900s, the anaesthetist was 'reading' patients, not as unpredictable constitutions, nor as universal, mechanistic systems, but as classifiable bodies. His skill lay in suspending their sensibilities during surgery, and successfully integrating them as a whole at the end of the process.

In Scotland, Europe and America, anaesthesia had retained the status of a manual task, and administrators remained subordinate to surgeons. But over the first decades of the twentieth century, specialist appointments with responsibility for teaching anaesthetics to students were made in all these communities. The English specialists had blazed a trail that other anaesthetists were keen to follow. 'Nothing is more ridiculous than the diffidence with which he [the administrator] accepts a subordinate role in the operation. He is just as important a personage as the operating surgeon, for the patient's life depends upon him just as much as it does upon the latter', wrote Berne practitioner F. L. Dumont in 1903.[130] For all the tensions of specialism, English anaesthetists had created a symbiosis with surgeons that became an aspiration of their fellows worldwide.

Anaesthetists were successful in establishing specialist anaesthesia by 1900 because they aligned their claims to appeal to the particular characteristics of English medicine. Rather than prioritising technological expertise that may well have laid open the possibility of constructing anaesthesia as an applied science which could be taught in the same manner as laboratory science, specialists emphasised the pivotal contribution of clinical skills in safeguarding anaesthetic safety.

Patients were safeguarded, not by a practical knowledge of gases and machines, but by the anaesthetist's 'incommunicable' knowledge which he had gained from clinical experience.[131] Anaesthetists capitalised on the willingness of many surgeons to concede responsibility for patient safety under anaesthesia and sought to establish themselves as authorities with jurisdiction over other doctors in matters of breathing gas. It was indeed a replication of the process undertaken across Victorian medical practice which had served to establish doctors as experts over the society's health and disease.

# Conclusion: The History of Anaesthesia

Twenty-first century anaesthesia with its sophisticated and 'high-tech' repertoire of drugs and techniques appears worlds away from the early use of ether and chloroform. Many of the dangers that Snow and the later specialists faced, particularly in regard to chloroform, have been obviated with the introduction of new gases such as halothane. Yet the purpose of the process – to save patients the pain of surgery – has not changed, and although the risk to life from anaesthesia has diminished – anaesthetic-related mortality is around 1:100,000 administrations – it remains of pivotal concern to practitioners. The 'focus of training in anaesthesia is concerned with the avoidance of disasters', noted Professor Aitkenhead in 1997.[1] The history of anaesthesia in the 1900s and onwards is beyond the scope of this book. But this concluding chapter seeks to summarise what anaesthesia tells us about nineteenth-century medicine and to sketch out some of the historical concerns which persist in the anaesthesia of the twenty-first century.

## The 'discovery' of anaesthesia

There are many intriguing questions and paradoxes which emerge from the history of anaesthesia and this study has attempted to address some of them: the 'scientisation' of the Yankee dodge; British surgeons' dislike of ether; the toleration of the risks of chloroform; and the emergence of English specialist practice. My analyses of these areas have stressed the way in which the successful establishment of new medical practices is dependent upon the context in which they are placed, and is shaped by the specific social, and cultural concerns of that community. The story of anaesthesia underlines the way in which we, as historians, need to acknowledge and embrace such specificities in order to extend our understandings of the complex relationship between medicine and the society.

The discovery that inhaling ether could produce an insensible, but still living, body that was impervious to the pain of surgery was primarily serendipitous. Anaesthesia, which in humanitarian terms stands as one of the greatest of all nineteenth-century medical discoveries, came about as a result of American dentists trying to improve the returns from their practice. This runs counter to our instinctive expectations that anaesthesia was the outcome of a deliberate and rational search for a solution to the problem of surgical pain, and indeed to what I have said here about medicine c.1840. Given the emphasis in the 1840s on the new scientific medicine – its teaching in medical schools and promotion in medical journals – we might expect anaesthesia to emerge from this way of thinking. We saw earlier how willingly doctors and patients in 1846 branded ether as a 'scientific' and 'rational' technique, and immediately distinguished it from the mysterious and inexplicable practices of mesmerism. For this reason most historical accounts have reconstructed this discovery so that it accords with the assumption that anaesthesia was a product of the new scientific medicine. But this is wrong.

Men like Morton and Wells were not using the new anatomical and physiological constructions of the body as building blocks in their search for a means to palliate dental pain. Their choice of nitrous oxide and ether grew out their experience and knowledge of these chemicals as recreational agents, as indeed did Long's experiments with ether anaesthesia. Hickman's work was different and seems to have been a direct response to the new understandings of the nervous system and its functions, and the process of death. But his contemporaries believed his method of deliberately creating a state of asphyxia was too dangerous to contemplate. The response of doctors to ether, 20 years later in 1846, confirms this view. As we saw earlier, medical students in the 1840s were taught that inhaling ether and nitrous oxide beyond a state of intoxication was dangerous, and certainly too risky to undertake on patients. Although elite practitioners, like Snow, were able to interpret the effects of ether on the body by drawing on the physiological work of Flourens, Hall and others, which recognised that separation of sensibility and life was possible, most doctors perceived it as similar to poisoning or asphyxia. It is also significant that when French physiologist Flourens undertook animal experiments using chloroform, in advance of Simpson's 'discovery', he rejected the agent as a possible substitute for ether, on the grounds that it was too dangerous for humans.[2] Had medicine been reliant solely on the bodies of science for the development of new medical techniques, then anaesthesia, as we know it, may well not have emerged.

And yet, when Morton publicised his Yankee dodge, elite practitioners, like Snow, found that they could place and rationalise the process within the new anatomical and physiological frameworks, and believed it to extend their understanding of these bodies of knowledge.[3] Whereas mesmeric anaesthesia had failed to gain acceptance in elite circles, the technique of inhaling ether was swiftly found to be at one with the new scientific medicine. We cannot ignore the way in which anaesthesia was 'scientised' by Snow and others in a form which has proved lasting and meaningful to twenty-first-century medicine. So we need to read the 'discovery' of anaesthesia as a paradox of serendipity and science which reveals how breakthroughs and innovations in nineteenth-century medical practice often came about by chance, and also acknowledges the way in which progressive doctors in the 1840s turned to science to establish new medical practices.

## The humanising of surgery

Historians have commonly assumed that the introduction of anaesthesia was swift and troublefree, and presented it as a discovery that improved patient experience of surgery but did not otherwise change surgery until it was combined with the antisepsis practices of the 1870s. This study, however, has shown that such views need re-evaluation. We have seen how anaesthesia became established in two broad stages: selective from 1846 to the 1860s, a time when its use had to be justified; and the shift to universal use from the 1860s to 1900 and beyond. It has become clear that the availability of pain-relief brought about fundamental shifts in surgical thinking and practice.

One of the strongest, if not surprising, findings of this study has been the extent to which the lack of an efficacious method of surgical pain-relief prior to 1846 had inhibited surgeons from using operations as a firstline treatment, and caused many patients to refuse surgery. In Britain, a succession of cultural and medical shifts from the mid-eighteenth century onwards combined to form a context in which surgical pain became increasingly problematic. By the 1840s one of the primary duties of doctors was to relieve pain and suffering. But surgical pain remained beyond medical control. The very low numbers of operations performed prior to 1846 did not reflect a lack of surgical expertise, nor patient antipathy to the cure or relief of disease through surgery, but rather the fact that most surgeons and patients viewed surgery as a last resort. The benefits of operations were in most cases unarguable, yet the fear of pain meant that many patients chose to tolerate their condition in preference to undergoing surgery.

We have seen that surgeons did not initially change their calculus of risk on the basis of ether – but patients did. The knowledge that their operation would be performed using a proven method of pain-relief caused many patients to agree to surgery that they had previously dismissed. And, even when chloroform became established as a high-risk agent, and details of fatalities were published widely in the general press patients still chose to trade the risks of chloroform against the benefits of pain-relief. Surgeons were far more cautious and anaesthesia remained a selective practice until the 1860s because for most surgeons, the ethical imperative was to preserve life, rather than prevent the pain of the experience. The shift to a universal use of anaesthesia during the last decades of the nineteenth century marked a revision of this calculus and occurred because of a change in surgical understandings of risk and pain.

In the 1840s, Snow's belief that anaesthesia acted as a physiological protective against the shock and risks of surgery, in other words that pain served no physiological function, was radical. As we have seen, most doctors constructed ether and chloroform as therapies which, although beneficial, added to the danger of operations. But the indisputable evidence provided by the countless operations on insensible patients during the first 20 years of anaesthesia, including some of the most severely injured patients on the battlefields of the Crimea, had strengthened the view that in the surgical context, pain and suffering were purposeless and often dangerous. By the 1860s the rationale for selective anaesthesia had diminished and most surgeons accepted that they had a primary duty to protect patients from the pain of surgery, even though the risk of fatalities remained. One of the consequences of this shift was that surgeons were stimulated to take a more radical approach to the employment of operations.

During the 1850s the overall numbers of operations had risen although a large proportion of these remained minor, rather than major procedures. By 1865 William Fergusson was calling for a revision of the 'conservative' approach of the previous 50 years which was based on an implicit trust in Nature's restorative powers. 'Nature's ways' were well and good, he said, only as long as matters were progressing in the right direction: 'non-interference equally displays the absence of good surgery'. Fergusson noted necrosed bone as one of the most common examples of delayed intervention; left in the body it developed into a 'most offensive and distressing occupant'. He urged surgeons to act promptly and operate to remove the infected bone. Surgeons, he claimed, were also prone to demonstrate 'needless delay' in the treatment of tumours and he polarised the treatment of two cases of tumour in the muscle of the lower leg. In the first, the disease was left to run its course, resulting in a 'monster aspect' – 'even amputation was of no avail' and the patient died. The second case was treated by radical surgery: 'both limb and life were saved by early local removal'. Fergusson attributed this new approach to: 'an improved skill in surgery, founded on scientific and practical data'.[4] His account does indeed accord with those of historians who have explained the shift to a more radical surgery which came through the emergence of the new cellular pathology of the 1860s – in particular, the work of Rudolf Virchow;[5] but this was only part of the story.

Although anaesthesia had brought the practical benefits of insensibility, the willingness of surgeons to re-evaluate the place of operations within surgical practice was highly dependent upon the way in which anaesthesia had humanised surgical work. Through the use of anaesthesia, surgeons were able to realise their capabilities to cut and correct infected flesh or damaged limbs without infringing humanitarian principles. As a consequence, they were free to reconstruct their culture and identity in a manner which stressed the pertinence of surgical practice to wider Victorian aspirations of the creation of a humane and civilized society.

Lawrence has shown how surgeons have 'shaped and reshaped' surgery's history over time in order to suit particular purposes.[6] From the 1850s onwards, there is evidence that surgeons distinguished their practice from that of earlier times by juxtaposing humanity and skill against coarseness and brutality. The 'great change' that had come over surgery, noted by Fergusson in 1852, focused on the way in which anaesthesia removed the need for surgeons to inflict pain on patients.[7] We saw earlier how surgical pain-relief encouraged surgeons to perform more operations on vulnerable patients, specifically women and children. Pernick has alluded to the way in which anaesthesia seemed to open up surgical practice to women doctors, and men who may previously have baulked at the psychological difficulties of operating on conscious patients.[8] Pre-anaesthetic surgery became characterised as a brutal practice; but the new surgery had become work fit for the 'highest minds' of society and a key participant in the march of progress: 'every limb saved is a blot removed from the face of nature's noblest work, and is a help to man's march along the steep and rugged road to heaven', explained Cambridge surgeon G. M. Humphry in 1864.[9] Similar rhetoric can be found in an 1867 dental advertisement for the performance of operations 'without pain' through the use of an anaesthetic spray: 'to relieve the bodily anguish of humanity is the noblest undertaking that can engage the attention'.[10] Thus by the end of the nineteenth century, London surgeon, Frederick Treves, could confidently dismiss the practices and attitudes of earlier generations:

> Treatment was very rough. The surgeon was rough. He had inherited that attitude from the days when operations were carried through without anaesthetics, and when he had need to be rough, strong and quick, as well as very indifferent to pain. Pain was with him a thing that had to be. It was a regrettable feature of disease. It had to be submitted to. At the present day pain is a thing that has not to be. It has to be relieved and not to be merely endured.[11]

Treves' words also encapsulate the way in which anaesthesia had redefined surgical responsibilities for pain. It seems that once anaesthesia was proven to create insensibility to the pain of surgery, surgeons found it increasingly difficult to justify suffering in other areas of their practice. Treatments and methods began to change in order to meet these new philosophies: disease of the hip-joint was treated by 'perfect rest', rather than the traditional use of blistering or poultices,[12] and a new gentleness appeared in the treatment of wounds and injuries. Humphry saw these shifts as fitting for the new humanitarianism of surgery: the 'anaesthetic treatment of wounds is the proper sequel to the anaesthetic mode of performing operations'.[13] In the surgical culture created by anaesthesia, pain management had become a new responsibility. Nor were the humanitarian influences of anaesthesia confined to medicine.

## The morality of pain

Mapping the social and cultural consequences of new medical practices is tricky, but we can use the notion of anaesthesia as a wider humanising force to explain shifts in two key late nineteenth-century debates: the anti-vivisection movement and religious justification of physical suffering.[14] Pain lay at the heart of each debate: the infliction of pain on animals was the focus of anti-vivisectionists, Christian theologians sought to account for the purpose of physical pain and suffering within a religious context. Although these concerns had arisen well before the advent of anaesthesia and can be understood as part of the wider humanitarian legacy of Enlightenment philosophies, it seems that by disproving the inevitability of pain anaesthesia brought new life to old concerns. Let us begin with the anti-vivisection movement.

From the mid-eighteenth century there were individuals who objected strongly to cruelty against animals, and the Society for the Prevention of Cruelty to Animals was founded in 1824. We saw earlier how Hickman limited his experimentation on animals, and the controversy engendered by Magendie's public demonstration of vivisection in London in the mid-1820s. Indeed, the swift extension of anaesthesia for operations on animals is indicative of this concern.[15] However, in the early 1860s, the period during which anaesthesia was shifting from a selective to a universal technique, an active and sustained campaign began to restrict medical and veterinary vivisection.[16] At this point, French notes, the arguments of the Royal Society for the Prevention of Cruelty to Animals (RSPCA) began to crystallise upon the specific cruelty of the pain of vivisection. Practitioners like Benjamin Ward Richardson, who used animal experiments to research the anaesthetic process, supported the campaign and proposed that anaesthesia itself could be the solution to the controversies of vivisection. If experiments were performed under anaesthesia, animals suffered no pain. If no pain was inflicted then charges of cruelty could not be sustained.[17] It was indeed a recapitulation of the arguments used by surgeons to promote their practice as humanitarian.

Paradoxically, although anaesthesia became to a large extent the means through which vivisectionists gained public acceptance of their practices, the new humanitarianism of surgery became a pivotal, although unacknowledged touchstone for the arguments of anti-vivisectionists. The 1876 Cruelty to Animals Act, for example, banned the use of vivisection for teaching purposes in response to claims that inflicting pain on animals could corrupt those who performed such acts.[18] Lewis Carroll expressed the horror of: 'successive generations of students, trained from their earliest years to the repression of all human sympathies, [who] shall have developed a new and more hideous Frankenstein – a soulless being to whom science shall be all in all'.[19] Such views could only have gained ground in the context of painless surgery; pushing for the protection of student sensibilities when all operations were carried out on feeling bodies would have been unsupportable. It is possible to sketch out similar shifts in relation to the religious debates on the purpose of pain which erupted during the late nineteenth century.

In literal interpretations of the Bible, argued theologians, eternal damnation centred on physical suffering through flames, torment and devils. Pain was unpleasant but just, and of ultimate benefit. Though social tolerance of physical pain had decreased from the mid-eighteenth century, and the doctor, rather than the priest, had become the prime authority at the deathbed, many believers continued to understand physical pain as a means of achieving divine salvation.[20] That medical justification of the physiological purpose of pain remained strong is illustrated by the selective use of anaesthesia during the 1850s. But around the 1860s with the extended use of anaesthesia, the new views of pain as purposeless, which had emerged in the 1820s, became dominant. And around this time, the debates in the Church of England over the meanings of bodily pain reached a new intensity.

Bending suggests that once pain had lost its medical rationale theologians found it increasingly difficult to retain public appreciation of bodily pain as just and necessary.[21] Reconciling physical suffering with the concept of a loving God increasingly became a stumbling block to belief in Christianity. From the mid-1860s many sermons addressed the problem of pain,[22] and the Church of England brought several clergymen to trial on charges of heresy. Defendants like Charles Voysey were dismissed from the Church for refusing to accept the idea of eternal damnation with its stress on physical pain and punishment.[23] Nor was the controversy limited to clergymen. Frances Power Cobbe, who spearheaded the anti-vivisection movement, was equally repulsed by the Christian portrayal of eternal damnation.[24] Bending shows how by the end of the nineteenth century there had been a significant change in emphasis: the physical pain of eternal damnation had been reconstructed as the spiritual pain of separation from God.[25] During this period it also appears that public fear of the pain of death increased and doctors stressed their ability to make death easy and comfortable.[26] In 1887 William Munk published *Euthanasia: or Medical Treatment in Aid of an Easy Death*, which became the definitive text on the care of the dying. Munk defined euthanasia in classical terms as 'a calm and easy death' and sought to reverse popular beliefs that the last stages of death were highly painful. Pain was no longer integral to medicine, or to redemption.[27]

Concern about the morality of inflicting pain spread to many areas of Victorian life. Corporal punishment in prisons and schools, for example, became a matter of public concern: the Whipping Act of 1861 began a 20-year period of legislative measures to control such practices; public executions ceased in 1868.[28] Wiener suggests that the shift away from using pain as a means of social control was largely due to the way in which anaesthesia had removed the rationale for pain.[29] By the end of the nineteenth century, doctors drew on anaesthesia as a prime exemplar of both medical humanitarianism and medical science. There is little difficulty in accepting the claims regarding the humanitarian benefits of anaesthesia, but evaluating its relations with the new medical science is far more problematic.

## Anaesthesia, science and practice

We have seen how from the beginning, anaesthesia was claimed as a star example of the benefits of the new medical science. But despite the rapid 'scientisation' of ether by would-be specialists, few doctors sought to determine their practice through chemistry and physiology. Instead, they employed the older principles of therapeutic practice which were based on the idea of a body as a holistic system. Simpson, of course, is a prime example of this approach, and we have argued that the different approaches of Snow and Simpson can be understood as exemplifying the different models of the body. The shift to a universal use of anaesthesia during the 1860s might suggest that it was around this time that Snow's view of anaesthesia as a universal process, suitable for all human bodies, gained acceptance. Yet it would be wrong to deduce from this that the practice of anaesthesia took on scientific characteristics that were typified by the new stress on determinism and predictability emerging from experimental physiology in the 1860s. Instead it remained an empirically driven practice throughout the nineteenth century and despite the emergence of specialists, unpredictable death remained an inherent risk of insensibility. To understand why this was so, we need to take account of the difficulties doctors encountered in trying to realise their aspirations that science determined practice.

That Snow's practice of anaesthesia was based upon the scientific principles he had established through experimental work is evident. But despite his exemplary track-record, it is clear that science did not provide the full security he hoped for. Experimental work, for example, failed to alert him to the particular dangers of amylene, from which he suffered two fatalities. Nor did it provide a satisfactory resolution to the debate over the mode of chloroform death.

Snow's experiments to establish the physiological mechanism of chloroform death led him to conclude that in cases of sudden death the heart was poisoned by too large a dose of the gas. For this reason he used inhalers that allowed controlled dosage of the agent. But as the use of anaesthesia grew during the 1850s, Snow's claim that the risks of chloroform could be removed through the use of inhalers failed to mesh with the experience of many doctors. Snow may well have used experimental science to pursue the anomalies in chloroform death further, had he not died unexpectedly in 1858. By the time the Chloroform Committee of the Royal Chirurgical and Medical Society published its report into anaesthetic agents in 1864, it was apparent that many of the deaths were not overdose cases. The Committee concluded that the method of administration was less significant than the clinical skill of the administrator.

After Snow's death no one continued his mission of promoting scientific principles as the basis for the practice of anaesthesia. There were several skilled practitioners by the 1860s, of whom the most highly rated was Joseph Clover. However, although Clover gained a strong reputation for efficacious practice and the development of anaesthetic technology, he published less than Snow and did not champion anaesthetic matters through medical networks.

Chloroform death remained high on the agenda of medical journals and societies throughout the last half of the nineteenth century, but the discussions focused on empirical issues rather than science, a focus that cannot be explained through a lack of interest on the part of physiologists. John Burdon Sanderson was among several English physiologists who investigated the mechanism of nitrous oxide anaesthesia in the late 1860s, and in Paris, the French physiologist Claude Bernard delivered a series of lectures on experimental research into anaesthetics in 1869 and 1870. Like Snow, Bernard desired that physiology should feed into pathology and therapeutics in order to serve as a 'solid and definitive base for the practitioner'. However, he acknowledged that 'we are still a long way from such a state of affairs'.[30] What can we conclude therefore about the relations between science and the practice of anaesthesia?

The lack of integration between experimental physiology and anaesthetic practice can be explained at several levels. In general, the empirical practice of anaesthesia was sustained by the wider medical culture of the 1850s onwards, which still privileged bedside skills above laboratory science.[31] This approach lent general support to the idea that anaesthesia should become a specialised practice, founded on a set of clinical skills and supported by technical ability, rather than a technically driven procedure. This strategy of promoting anaesthesia as a high-risk practice, in which the safety of the patient depended upon the clinical experience and skills of the administrator, reinforced medical aspirations of authority and status. Creating an image of doctors who were willing to tolerate the pressures of performing such a risky procedure to save patients the pain of surgery, also underlined the humane aspects of medicine. In late nineteenth-century British medicine, these were vital measures of professional worth.

More specifically, anaesthetists believed that physiological experiments on animals were unable to comprehensively mimic the effects of anaesthesia on patients, in particular the power of the mind to influence the outcome of inhalations. We saw earlier how patient fear of surgery had metamorphosed into a fear of anaesthesia, specifically the experience of 'going unconscious'. In practice, fearful, nervous patients were known to be at far more risk of anaesthetic complications. Vivisections, said practitioner Charles Kidd, in 1867, were: 'erroneous and misleading... purely physiological, not surgical experiences, where disease, emotion, idiosyncrasy... are lost sight of'.[32] There were also differences of opinion between physiologists and anaesthetists in their interpretation of physiological signs. Frederic Hewitt described how whilst visiting a laboratory in the 1890s, a question arose as to the breathing of an animal under experiment. 'To my astonishment', he said, 'I saw that respiration was regarded as proceeding so long as certain feeble and fitful muscular contractions about the chest and abdomen persisted – contractions which were and for some time had been, utterly unable to cause the slightest ingress or egress of air.' It showed, he added, how a physiologist's view of respiration was mistaken, and that 'a practical acquaintance with anaesthetising human beings would materially assist the physiologist in his researches'.[33] From the anaesthetist's perspective, the waters of clinical practice were wider and murkier than laboratory findings seemed to suggest.

The divide between experimental physiology and anaesthetic practice continued to 1900 and beyond. Although the emerging London anaesthetists claimed that their practice of classifying patients into anaesthetic 'types' was making anaesthesia a more predictable and safer process, in many parts of the country the Scottish method of giving chloroform on a cloth remained in vogue. Indeed the method had been validated in 1890 by the 2nd Report of the Hyderabad Chloroform Commission, which concluded that respiration, rather than the heart, was the area of risk.[34] When experimental science did provide an explanation of sudden deaths under chloroform, through Levy's experiments of the 1910s which proved that chloroform could kill in low doses by inducing atrial fibrillation, there was little, if any, impact upon anaesthetic practice.[35] Chloroform retained its popularity and remained in use in Britain until the 1950s. The controversy over its risks, which had arisen in 1848 following the death of Hannah Greener, was eventually transcended, rather than resolved, by the introduction of new anaesthetic agents.

Thus the role played by science in the practice of anaesthesia emerges as one which runs contrary to current historical assumptions. During the 1840s, a time when many historians have understood science to play a rhetorical rather than practical role, we have evidence of experimental science being used as the foundation for practice through the example of Snow.[36] Conversely, it appears that during the last decades of the nineteenth century, a period when experimental physiology was beginning to mesh with many areas of medical practice, the practice of anaesthesia remained isolated from the laboratory and developed empirically. It lends weight to the view, emerging from new historical work, of the need for further evaluation of the variegated and diffuse characteristics of British nineteenth-century science and the relations it enjoyed with medicine.[37]

## Culture and diversity

We have learnt much during the course of this study about the different dynamics which governed relations between patients and doctors in the nineteenth century. Patient approbation of painless surgery is unquestionable; out of the 4500 or so anaesthetic administrations that Snow recorded, there were only 5 instances in which the patient either refused anaesthesia or the inhalation was aborted at the request of the patient. And our experience of living within a culture where few medical interventions are undergone without pain-relief makes it easy to understand why patients were so reluctant to consent to operations prior to ether. But the worldwide variation in practice during the first 50 years of anaesthesia is of particular interest to historians because it can best be explained through local influences of culture.

We have seen, for example, how within months of its introduction, the practices of ether in America and Britain had begun to diverge. The shift in Boston to administering ether on a cone, rather than an inhaler, was a direct consequence of Morton's attempt to patent his inhaler. However, the fact that surgeons persisted in this method, rather than switching to another form of inhaler, suggests that they were comfortable with its 'heroic' characteristics – notably the use of force to ensure patients became insensible as quickly as possible. It sat easily within the dynamics of American patient–doctor relationships. However, in Britain, surgical practice was more tempered. Surgeons sought patient cooperation and were reluctant to force patients to inhale; a strong view emerged that ether exacerbated, rather than eased, the difficulties of controlling patients during operations. It is possible to explain part of this set of circumstances at a material level – ether inhalers were inefficient and required patient cooperation in tolerating the pungent vapour and breathing through hard rubber tubes and facemasks in a way that a sponge or a cloth did not. But as earlier chapters have shown, the response of surgeons to ether only becomes fully meaningful when it is placed within the context of Victorian values of self-control and takes account of the way in which this cultural aspiration had fed into the surgical context.

The contrast between medical and social expectations of surgical practice in different communities is even more vivid in the varying responses to the risks of chloroform. We have seen how in some places, particularly the northern states of America and parts of Europe, doctors responded to the occurrence of deaths by abandoning chloroform and returning to ether. But British doctors and those in the southern states of America and most in Europe chose to accommodate the risks of chloroform. Doctors claimed that patients preferred chloroform because it was more pleasant to inhale, yet there seems little doubt that most patients would have preferred to tolerate ether rather than submit to operations without pain-relief. Although it is possible that private patients could have influenced the choice of anaesthetic, Chapter 5 showed how in London hospitals surgeons had autonomy over the use of anaesthesia, and chloroform remained the primary anaesthetic agent until the 1870s. We are led to conclude that the British use of chloroform stemmed from a surgical desire for its ease, which was sustained by a tolerance on the part of patients that doctors could not be held responsible for the risks of anaesthesia.

We have seen that the first inquest on an ether fatality ruled that as the use of anaesthesia during surgery was informed by the doctor's desire to remove the pain of the experience, he should be exonerated from any liability for its subsequent effects. Subsequent inquests continued this approach. Doctors were held accountable for administering anaesthesia within established principles – ensuring the patient received plenty of oxygen so as to avoid asphyxia and limiting the strength of the dose – but not for the fatal effects of the chemical. There is no evidence, for instance, that British patients challenged surgeons over their choice of chloroform compared to ether. Public debates about the safety of anaesthesia rumbled on well past 1900 but never resulted in government intervention, unlike the debates on public safety and poisons in the early 1850s, nor reached a crescendo of public hostility like the anti-vaccination debates of the 1870s. British patients' fear of pain caused them to endure the risks of chloroform and accept medical jurisdiction over the practice of anaesthesia. We suggested that such tolerance sat easily with the Victorian public's acceptance that many of the artefacts of progress – railways, industry and so on – contained innate elements of risk as well as benefits.

In the northern states of America patients had very different expectations of medical responsibility. The choice of safety above ease – ether over chloroform – can be understood as a direct response to the growing incidence of medical malpractice suits. From the 1830s onwards, surgical practice in these communities became increasingly subject to scrutiny and complaint, and this caused surgeons to exercise caution and adapt their practice to minimise all possible risks. It was this increasing emphasis on surgical accountability for clinical outcomes that caused surgeons to abandon chloroform in favour of ether.

That medical communities responded differently to the risks of chloroform again emphasises the intricate and complex relationship between the practice of medicine and its social and cultural context. The different strategies adopted by doctors to manage the risks of anaesthesia in medical practice can therefore be taken as a measure of the boundaries of medical and individual responsibility established in each community. In this way, it is possible to make some sense of the differences in anaesthetic practice at all levels – local, national and international. It is also useful to understand that such historical divergences have long-lasting repercussions.

## Continuities

So my aim in this concluding section is to show how the professional structures and practices of twenty-first-century anaesthesia continue to follow templates created in the nineteenth century. Nowhere is this more apparent than in the contrast between the current status quo of anaesthesia in Britain and America. Chapter 6 traced the emergence of specialist practice in England and showed how anaesthetists justified their claims by arguing that experience was the best means of eliminating risk. There is no doubt that specialists were highly receptive to new gases and methods, and technology grew to characterise specialist practice. But they also found it strategically wise to stress the uncertainties posed both by patient fear of anaesthesia and by the inherent unpredictability of the process, and to claim to minimise these through the development of a particular clinical routine. In this way they manufactured a practice centred on reducing risk and protecting patient safety; their success was made evident by the appointment of anaesthetists in many large English hospitals before 1900. Although until the inauguration of the National Health Service in 1948, anaesthetists did not achieve financial and academic parity with their surgeon and physician peers, and general practitioners undertook the majority of administrations, the early specialists secured clinical responsibility for patient safety during anaesthesia which indubitably confirmed the practice of administering gas as a medical role.[38] In Britain today anaesthesia remains exclusively under the authority of anaesthetists. However, America differs.

We have seen how patterns of anaesthetic use diverged across America through the last half of the nineteenth century: ether remained the primary agent in the northern states; in the south, chloroform retained popularity. But north and south replicated the Scottish pattern of using students or nurses to give anaesthesia, as indeed did most of Europe. A few centres in Europe appointed chloroformists during the 1880s, and the first American specialist appointment was made in 1897 in New York.[39] Over the first half of the twentieth century, America followed the English example of developing the practice of anaesthesia as a medical specialty, although it was not recognised by the American Medical Association until 1937.[40] However, although the new anaesthetists took on management and research responsibilities, nurse anaesthetists continued to administer many of the routine anaesthetics, and a three-tier system of practitioners evolved: anesthesiologists (physicians), nurse anaesthetists and anesthesiologist assistants.[41] Anesthesiologists are usually self-employed and contract their services to hospitals, often in a mixed team with nurse anaesthetists and anesthesiologist assistants. A question mark thus hovers over the status of American anaesthesia: Is it a nursing or a medical role and how should responsibility be divided?

The 'considerable discord' between anesthesiologists and nurse anaesthetists is currently attracting much medical and public attention although relations between the different practitioners have been uneasy for many years.[42] Bunker has suggested that conflict arose after the Second World War, the point when anaesthesia became a significant medical specialty. The early anesthesiologists, he noted, constructed their relationship with nurse anaesthetists upon ideas of professional competition, rather than seeking collaborative practice.[43] Indeed the term 'anesthesiologist' was introduced as a means of distinguishing physician practice from that of nurses. Since the 1970s there have been a complex series of political and economic changes bringing the debate to a new pitch. For example, in the 1990s, medical insurance companies changed payment methods which had been in place since the 1960s and which had created highly lucrative arrangements for anesthesiologists. In 1993 anesthesiologists came second in the hierarchy of physician earnings, exceeded only by cardio-thoracic surgeons.[44] But the changed billing arrangements have reduced the profitability of anaesthesiology and as a consequence, the professional rivalry between anesthesiologists and nurse anaesthetists has increased. Currently, nurse anaesthetists deliver around 60 per cent of all anaesthetics and earn in the region of one-third of an anesthesiologist's salary.[45] In some states of America, nurse anaesthetists have sought to circumvent these pressures by seeking independent practice rights, thus placing themselves in direct competition with anesthesiologists. Both parties employ political lobbying and a wide range of media to pursue their cause, and one of the most interesting aspects of this confrontation is the fact that both parties mount their campaigns on issues of safety and risk. In this they replicate the arguments of the early English specialists.

It is a moot point whether late nineteenth-century specialist practice saved deaths under anaesthesia. It remains so in the early twenty-first century. Although American anesthesiologists have sought to distinguish their practice from that of nurse anaesthetists by linking it to better outcomes, there has been to date no definitive substantiation of this claim.[46] Safety is also the key issue for nurse anaesthetists. Publicity material produced by the American Association of Nurse Anesthetists bears the slogan 'Nurse anaesthesia = Safe anaesthesia'. Another poster shows a scene in an operating theatre in which a masked figure towers over a prostrate, unconscious patient. It opposes health insurance company Tricare's proposal for allowing anaesthetics to be given by anesthesiologist assistants. The text threatens that if this is accepted: 'The person putting you under for this surgery could be an *assistant.*' 'Call your Representative and Senator . . . and say NO to Anesthesiologist Assistants.'[47] Patient fear of anaesthesia also remains part of the debate and in both camps the provision of a caring and empathetic service forms a major strand in the defence of their claims to best practice. Avoiding pre-operative anxiety was one of the most important factors listed by anesthesiologists in a 1999 study.[48] Nurse anaesthetists have claimed that patients prefer their type of care.[49]

It is no surprise then that American anesthesiologists envy the status and autonomy enjoyed by their British counterparts. But political and economic pressures such as a shortage of anaesthetists and the introduction of the European Working Time Directive mean that British anaesthetists are facing the possibility that nurse anaesthetists will have to be incorporated into their structures of practice. Professional associations have so far resisted such measures and have used the American situation as a worst-case scenario. In 2003, the UK NHS Service Delivery and Organisation Research and Development Programme commissioned a study on the options facing anaesthetic services. The findings of the draft report, *Exploring Professional Boundaries in Anaesthesia*, are optimistic that nurse anaesthetists could be introduced to British practice without creating antagonism between providers. Nevertheless, they emphasise the complexities of introducing a new role into an established practice and suggest that concerns about risk and safety may to some extent be masking unexpressed anxieties such as job security and professional status. We have yet to see how these current challenges to British anaesthesia will be resolved. What does seem certain is that in Britain, anaesthesia will remain securely within the medical domain. In their response to the study, the Association of Anaesthetists of Great Britain and Ireland defined the status of anaesthesia as a 'medical act [which] requires supervision by a medically qualified anaesthetist'.[50] To ensure that patient safety is made paramount, the Association requires that any new role should be placed 'always under the supervision of a medically qualified anaesthetist'.[51] The tenets of modern practice thus remain consistent with the philosophies of the early specialists.

The emergence of specialist anaesthesia in England during the late nineteenth century was neither random nor inevitable. It was a product of local culture: the influence of Snow in establishing technology as a method of safety; Victorian society's aspirations of control which were so easily fulfilled by chloroform; the dynamics of elite London practice and the stress on 'bedside' medicine; and the willingness of aspiring anaesthetists to bear responsibility for administering gas, in return for professional and economic benefits. Whatever the future holds for British anaesthetists, it seems likely that they will continue to derive benefit from the legacy of a particular historical moment.

# Appendix

*Table A.1* London operations under ether, by sex, January–March 1847

| | No. of patients |
|---|---|
| Males | 52 |
| Females | 25 |
| Total | 77 |

*Sources*: *L, LMG, MT*, Robinson 1983.

*Table A.2* London operations under ether, by age, January–March 1847

| | 0–10 | 11–20 | 21–49 | 50 upwards |
|---|---|---|---|---|
| No. of patients | 8 | 8 | 48 | 10 |

*Sources*: *L, LMG, MT*, Robinson 1983.

*Table A.3* London operations under ether, by type of procedure, January–March 1847

| Type of procedure | |
|---|---|
| Amputations Major | 22 |
| Amputations Minor | 7 |
| Bladder/urethra | 4 |
| Bones/joints | 3 |
| Plastic/superficial | 21 |
| Tumours | 6 |
| Hernia | 2 |
| Eyes | 1 |
| Dental | 10 |
| Caesarian section | 1 |
| Total | 77 |

*Sources*: *L, LMG, MT*, Robinson 1983.

*Table A.4* King's College, 1845–60, Increase in Fergusson's rate of surgery

| Date | Number of surgical admissions | Number of patients operated on | % patients operated on | Increase in rate of surgery |
|---|---|---|---|---|
| 1845–46 | 138 | 17 | 12 | |
| 1847 | 117 | 37 | 32 | 2.6 times |
| 1859–60 | 70 | 38 | 54 | 4.5 times |

*Source*: KH/CN/68–84, King's College Hospital.

*Table A.5* King's College, 1845–60, Increase in Fergusson's rate of surgery, by type of procedure

| Type of procedure | 1845–46 | | 1847 | | 1859–60 | |
|---|---|---|---|---|---|---|
| | Number of operations | % total | Number of operations | % total | Number of operations | % total |
| Amputation major | 2 | 12 | 2 | 5 | 3 | 8 |
| Bladder/urethra | 4 | 23 | 9 | 26 | 12 | 31 |
| Bones/joints | 2 | 12 | 7 | 20 | 7 | 18 |
| Plastic/superficial | 6 | 35 | 8 | 23 | 8 | 21 |
| Tumours | 2 | 12 | 3 | 8 | 7 | 18 |
| Hernia | 1 | 6 | 1 | 3 | 1 | 3 |
| Other | | | 6 | 17 | | |
| Total | 17 | | 36 | | 38 | |

*Sources*: KH/CN/68–84, King's College Hospital.

*Table A.6* King's College, 1847, Fergusson's use of ether, by sex of patient

| | Operations with ether | Operations without ether | % with ether |
|---|---|---|---|
| Males | 8 | 19 | 30 |
| Females | 4 | 5 | 44 |
| Total | 12 | 24 | 33 |

*Sources*: KH/CN/68–84, King's College Hospital.

*Table A.7* King's College, 1847, Increase in Fergusson's rate of surgery for male patients

| Date | Number of males admitted | Number of males operated on | % males operated on | Increase in rate of surgery |
|---|---|---|---|---|
| 1845–46 | 78 | 12 | 15 | |
| 1847 | 67 | 24 | 36 | 2.4 times |
| 1859–60 | 49 | 22 | 45 | 3 times |

*Sources*: KH/CN/68–84, King's College Hospital.

*Table A.8* King's College, 1847, Increase in Fergusson's rate of surgery for female patients

| Date | No. of females admitted | No. of females operated on | % females operated on | Increase in rate of surgery |
|---|---|---|---|---|
| 1845–46 | 60 | 5 | 8 | |
| 1847 | 50 | 9 | 18 | 2.2 times |
| 1859–60 | 49 | 22 | 45 | 5.6 times |

*Sources*: KH/CN/68–84, King's College Hospital.

*Table A.9* London chloroform fatalities, 1849–57

| | 1849 | 1850 | 1851 | 1852 | 1853 | 1854 | 1855 | 1856 | 1857 |
|---|---|---|---|---|---|---|---|---|---|
| No. of fatalities | 1 | 1 | 2 | 1 | 3 | 4 | 1 | 1 | 1 |

*Source*: OC.

*Table A.10* London chloroform fatalities, 1849–57

| Date | Location | Surgeon | Operation | Method |
|---|---|---|---|---|
| October 1849 | St Thomas' | Mr Solly | Toenail removal | Chloroform in inhaler |
| June 1850 | Guy's | Mr Cock | Part of hand | Chloroform on napkin |
| April 1851 | Stepney Workhouse | Unnamed | Penis | N/A |
| July 1851 | Seaman's Hospital | Unnamed | Removal of testicle | Chloroform on lint |
| March 1852 | St Bartholomew's | Mr Lloyd | Aneurism | Chloroform in inhaler |

| | | | | |
|---|---|---|---|---|
| March 1853 | University College | Unnamed dresser | Ulcer on vagina | Chloroform on lint |
| October 1853 | University College | Dr Hillier | Strangulated hernia | Chloroform on lint |
| October 1853 | St Bartholomew's | Mr Paget | Cautery to vagina | Chloroform in inhaler |
| May 1854 | Lock Hospital | Unnamed | Phymosis | Chloroform in inhaler |
| July 1854 | Middlesex Hospital | Mr De Morgan | Amputation of thigh | Chloroform in inhaler (Snow's) |
| October 1854 | University College | Mr Erichsen | Catheter insertion | Chloroform on lint under towel |
| December 1854 | Guy's | Mr Bryant | Amputation of leg | Chloroform on lint |
| April 1855 | Royal Ophthalmic Hospital | Mr Bowman | Excision of eyeball | Chloroform in inhaler (Snow's) |
| October 1856 | St Thomas' | Unnamed | Finger removed | Chloroform on lint |
| August 1857 | King's | Mr Heath, house surgeon | Syphilitic warts | Chloroform in inhaler (Snow's) |

*Source*: *OC*.

*Table A.11* Snow's Obstetric Practice, 1849–57

| | **No. of labours** | **Attended as GP** | **Used chloroform** | **Attended as specialist** |
|---|---|---|---|---|
| 1849 | 24 | 15 | 2 | 9 |
| 1850 | 15 | 14 | 2 | 1 |
| 1851 | 17 | 13 | 0 | 4 |
| 1852 | 7 | 3 | 0 | 4 |
| 1853 | 20 | 5 | 1 | 15 |
| 1854 | 12 | 5 | 0 | 7 |
| 1855 | 15 | 3 | 1 | 12 |
| 1856 | 17 | 4 | 0 | 13 |
| 1857 | 16 | 0 | 0 | 16 |

*Source*: *CB*.

# Notes

## Introduction

1. *OE* 56–7.
2. *MTG* 15 (1847) 274.
3. See for example Bynum and Porter 1993, Loudon 1997, Porter 1999.
4. Bergman 1998. See also Caton 1999, Maltby 2002, Rushman 1996.
5. Bryn Thomas 1975, Cartwright 1952, Keys 1945, Smith 1982, Sykes 1992.
6. Many contemporary textbooks of anaesthesia contain introductory chapters charting the origins of practice. For example, Prys-Roberts and Brown 1996.
7. Duncum's account of the development of anaesthetic technology remains the definitive work. Duncum 1994.
8. Pernick 1985.
9. Notable studies by historians are Burney 2006a, Lawrence 1992, Poovey 1987, Winter 1991, 1998.
10. Lawrence 1987, p. 12.
11. Ackerknecht introduced the contrast between bedside medicine and hospital medicine. Jewson elaborated the biographical aspect of bedside medicine, and linked it with the patronage of physicians by high-class patients; he also developed the contrast between hospital medicine and the later medical work done in laboratories. Pickstone suggests that biographical medicine might in part be seen as a kind of natural history of patients or diseases, and also as involving cultural understandings by doctors and patients through which, for example, diseases are understood in terms of morality or destiny. He sees clinical examinations, post-mortems and most of the work of hospital laboratories as kinds of *analysis*, and he contrasts this kind of science with the *experimentalism* of the later nineteenth century, which manipulated living animals in the hope of gaining mastery over medical phenomena. He has also stressed that these types should not be seen as simply successive, but as cumulating in complex and contested ways. Ackernecht 1967, Jewson 1976, Pickstone 1993, 2000.
12. See Pickstone 2000, pp. 10–12 and Chapter 5.
13. See for example Shortt 1983, Warner 1980.
14. Snow 1995. See also Bynum 1994.
15. Snow's original casebooks are held by the Royal College of Physicians, London and a transcription was published in 1994 by the late Richard Ellis. On the wealth of information contained in the casebooks, see Warner 1996.
16. Porter 1985.
17. For example Digby 1994, Lawrence 1994, Peterson 1978.
18. On specialisation see Friedson 1970, Lomax 1996, Rosen 1972, Stevens 1966.
19. But see Cooter 1993, Davidson 1996 and Rosen 1972.
20. For example see Granshaw 1989.
21. Granshaw 1992.

22. Cooter 1993.
23. Davidson 1996, Rosen 1972.
24. On eighteenth-century dentistry, see Hillam 2003. The way in which the mouth and teeth became the focus of a new body of knowledge in the mid-nineteenth century is considered in Nettleton 1988.
25. Tomes published a series of lectures in 1847 on *Dental Physiology and Surgery* and was a founder member of the Odontological Society in 1856 and the College of Dentists in 1857. Both of these organisations were influential in establishing the Dental Hospital and London School of Dentistry, which opened in Soho Square in November 1858. Later, Tomes was a founder member of the British Dental Association. See Cohen 2004, Lufkin 1938.
26. See letter from James Syme arguing that the shock of pain was a more dangerous depressant than chloroform, *Times*, 12 October 1854, page 9, column a.
27. See Shepherd 1991, Connor 1998.
28. Snow expressed concern that the use of the term *anaesthesia* should not cloud the true chemical processes of ether and chloroform – narcotism – and thereby suggest that their physiological action differed from other chemicals in that group. He also noted that anaesthesia may well convey the impression that unconsciousness was part of the phenomenon whereas the privation of feeling – the true meaning of anaesthesia – was not necessarily accompanied by unconsciousness. *OC* 34 and see Duncum 1994, Appendix B.

## 1 Enlightenment philosophies

1. Bergman 1998. See also a review of Bergman by Douglas Bacon who describes his lack of speculation to be the 'one great failing of the book'. *Bulletin of the History of Medicine*, 73 (1999) 319–20.
2. Cartwright 1952, pp. 328–9.
3. Smith 1982, p. 38.
4. Jacob and Sauter 2002.
5. In the following section, I draw heavily on Bynum 1994, Lawrence 1994, Pickstone 2000, Porter 1999.
6. Bynum 1994, p. 14, Porter 1999, p. 250.
7. Bynum 1994, p. 14.
8. See Lawrence 1994, p. 15.
9. Porter 1990, p. 20.
10. Lawrence 1994, p. 15.
11. Fissell 1992.
12. Porter 1999, p. 262.
13. Bynum 1994, p. 12.
14. Lawrence 1994, p. 14.
15. Bynum 1994, p. 17.
16. After his time in Edinburgh, Rush went to America where his 'heroic' therapeutic approach had widespread impact. For a discussion of his influence upon therapeutic practice there, see Warner 1986, pp. 27–30. On the ways

in which Rush's 'heroic' approach influenced the American take-up of ether in 1846, see Pernick 1985, pp. 13–5.

17. On the wider influences of Brunonian medicine, see Bynum and Porter 1988.
18. Porter 1992.
19. Brown and Cullen disagreed on whether opium acted as a stimulant or depressant and the argument was extended by other doctors well into the nineteenth century. Hayter 1971, pp. 28–9, Berridge and Edwards 1987, pp. 64–6. The same dichotomy was later drawn between the properties of ether and chloroform, see *OE* 33.
20. Bynum 1994, p. 17.
21. Porter 1992, p. 13. Beddoes edited Brown's *Elements of Medicine*, see Hayter 1971, p. 27.
22. Porter 1992, p. 4.
23. Pickstone 2000, p. 89.
24. Quoted in Lawrence 1990, p. 214.
25. Uglow 2002.
26. On Davy and Beddoes, see Cartwright 1952, Golinski 1992, Knight 1992, Lawrence 1990, Porter 1992, Smith 1982, Stansfield 1984.
27. Quoted in Knight 1992, p. 24.
28. Ibid., p. 25.
29. Davy 1972, p. 334.
30. Ibid., p. 335.
31. Ibid., p. 466.
32. Ibid., p. 335.
33. Ibid., p. 342.
34. Ibid., pp. 347, 449.
35. Quoted Porter 2003, p. 215. See also Aries 1991, pp. 397–401, Albury 1993, Stevenson 1979.
36. Curry 1792, pp. v–x. The 'newness' of the view is reinforced by Edmund Goodwyn's definition of the 'two general conditions' of the animal body in 1788, as 'Life and Death'. 'By death', he wrote,

    > we understand the privation of life, there can be no intermediate state between them. Of the body in this disease, we can say with propriety only, that it is alive or that it is dead. If it were really dead, it would necessarily follow, that the means which are employed to recover it...must be supposed to communicate life to dead matter, which is impossible.

    He criticises the use of terms such as 'suspended life' or 'suspended animation' on the grounds that they 'lead mankind to believe themselves capable of reanimating or resuscitating a lifeless man, when they only cure a disease'. Goodwyn 1788, p. 99. Davy and Beddoes were familiar with Goodwyn's work, as was Hickman. See Cartwright 1952, pp. 273–4.
37. Johnson 1773.
38. Bishop 1974.
39. Jackson 1996, pp. 102–103.
40. Davy 1972, p. 555.

41. Smith 1982, pp. 11–17.
42. Cartwright 1952, pp. 97–9.
43. Davy 1972, p. 458.
44. Ibid., pp. 460, 462, 488.
45. Quoted in Hayter 1971, p. 75.
46. Davy 1972, p. 516.
47. Lawrence 1990, p. 218.
48. Davy 1972, p. 556.
49. Ibid., pp. 464–5.
50. Davy 1972, p. 556. Davy's conception of nitrous oxide as a stimulant which increased blood flow explains his caution regarding operations with great blood loss.
51. Davy 1972, p. 552.
52. Ibid., p. 471
53. Written by Davy on 21 November 1800, quoted in Jacob and Sauter 2002, p. 164.
54. Golinski 1992, pp. 176–87.
55. Porter 1992, p. 49.
56. Yalton 1984.
57. Golinski 1992, p. 170.
58. Golinski 1992 and Fara 1996.
59. Fara 1996, p. 130.
60. Ibid.
61. Porter 1992, p. 19. Golinski's argument that the work of the Pneumatic Institute 'degenerated into a fiasco' at the turn of the century, primarily because of its shared identity with radical Enlightenment science, is rejected by Jacob and Sauter as offering 'too much power to the forces of reaction', but they in turn fail to take account of the way in which Davy's move to the Royal Institution marked a discernible shift in his approach to chemistry. Golinski 1992, particularly Chapter 6, Jacob and Sauter 2002, p. 175.
62. Berman notes the mesh between science, agriculture and philanthropy in 1799 was very different to that of later years; by the 1830s, science as avocation had become science as enterprise. Berman 1972, p. 229. The emergence of the 'gentleman' scientist as expert has been well-explored by historians and the pattern of knowledge-transfer at the Royal Institution shared the characteristics promoted by the British Association for the Advancement of Science from the 1830s. See Morrell and Thackray 1981 and Rudwick 1985.
63. Knight 1992.
64. Golinski 1992, Chapter 7.
65. Beddoes anticipated apparatus for breathing 'different kinds of air' would become part of all households; Davy suggested a method for producing large quantities of nitrous oxide economically so that the cost would not prohibit its widespread use. Golinski notes that the diffusion of new knowledge amongst the public had also been an aim of Joseph Priestley. See Golinski 1992, p. 8 and Chapter 6.
66. Quoted in Knight 1992, p. 42. Stansfield suggests that Davy made a deliberate attempt to distance himself from Beddoes and the Pneumatic Institution in

order to avoid the taint of political criticism. Stansfield 1984, p. 169. It is also possible that Davy was disillusioned by the unpredictability of his experiments and their failure to reinforce the therapeutic principles of Brunonian medicine. Davy 1972, pp. 557–8.

67. Austen 1961, pp. 6, 291.
68. Porter 1992, pp. 52–3.
69. Davy's retrospective view of nitrous oxide in 1829 was as a gas that was unfit to support life; although the experience of insensibility was 'agreeable', the process of recovery was painful. Jacob and Sauter 2003, p. 173.
70. One of the most recent formulations of this argument can be found in Papper: 'It took so long to "discover" anaesthesia because there was no societal readiness for it nor interest in the prevention of the pain of surgical intervention or really in the relief of pain for the common man until the idea that humankind had a right to "pursue happiness"'. Papper 1995, p. 22. See also: '[anaesthesia] is a dramatic outward expression of man's inner change from eighteenth century brutality to our own more human pattern of behaviour', Cartwright 1968, p. 33 See also Caton 1985, p. 498.
71. See Morris 1991, Rey 1995.
72. Porter 2003, p. 403.
73. The shift in attitudes to death was paralleled by marked change in the meanings attributed to suicide during this period. See Jackson 1996, Chapter 3, pp. 64–5.
74. Percival 1985, pp. 32, 39.
75. Quoted in Jalland 1996, p. 84.
76. Although evangelicism is acknowledged as a crucial force in 'shaping the mentality of the period', historians have not yet achieved a satisfactory evaluation of its many aspects. Boyd 1988, p. 7. See also Briggs 1979, pp. 66–74. On evangelicism and death, see Jalland 1996, Chapter 1.
77. Jalland 1996, p. 33.
78. *L* I (1838–39) 548.
79. Quoted in Jalland 1996, p. 88.
80. Porter suggests that the construction of pain as a 'beneficial reflex mechanism' was particularly appealing to advocates of Enlightenment philosophies. Porter 1993, p. 1581.
81. Rey 1995, p. 92.
82. Stanley 2003, pp. 53–6.
83. *L* II (1839–40) 564. See also Macartney 1838, review in *L* I (1838–39) 238–42, 268–71 and subsequent letter from Macartney in *L* II (1839–40) 562–64.
84. *L* I (1844–45) 72 and *L* II (1844–45) 246.
85. Witten 1991. See also Warner's discussion of the way in which the qualities of orthodox versus homeopathic medicine became polarised in American medicine as forceful/heroic versus timid/gentle. Warner 1998.
86. Parssinen 1983, pp. 23–4.
87. London physician Jonathan Peirera described opium 'the most important and valuable remedy of the whole Materia Medica'. Quoted in Parssinen 1983, pp. 22–3.

88. Berridge and Edwards 1987, p. 66.
89. Fissell 1991, Jacyna 1989.
90. Analyses of this shift can be found in Hannaway and La Berge 1998, Lesch 1984, Maulitz 1987, Pickstone 2000.
91. Nicolson 1993, Porter 1993, Reiser 1978.
92. Warner 1986, pp. 85–91.
93. Lawrence 1992, pp. 6–7.
94. On the Hunters, see Bynum and Porter 1985.
95. Richardson 1987.
96. Cooper 1822.
97. Stanley 2003, pp. 41–3.
98. James Moore, *A method of preventing or diminishing pain in several operations of surgery* (London, 1784), quoted in Druitt 1854, pp. 635–6. See also the example of Horatio Nelson being given opium after the amputation of his arm in 1797, rather than before, in Kirkup 1988.
99. Bivins 2000, p. 73 and Rey 1985, pp. 139–41.
100. *L* II (1832–33) 596.
101. *L* II (1832–33) 597. Rey notes how during the Napoleonic wars, Larrey, the chief surgeon, deliberately performed amputations swiftly in order to maximise the benefits of the state of 'stupor' which affected the injured part in the immediate aftermath of injury. Rey 1995, pp. 137–8.
102. Loudon 1986.
103. Hickman 1824. Suspended animation commanded much medical attention at this period, see for example Stowe 1823, Whiter 1819, Cartwright 1952, Stevenson 1979.
104. Berman 1972, pp. 224–5.
105. Carbon dioxide had recognised 'antiputrescent' qualities. See Cartwright 1952, p. 34.
106. *Gentleman's Magazine*, January–June (1825) 628, quoted in Cartwright 1952, p. 292.
107. During the German wars, Larrey, Napoleon's chief surgeon saw several cases of soldiers – suffocated by carbon dioxide fumes whilst sleeping in overheated rooms – who had been brought to hospital in a state of suspended animation from which several had recovered. He noted that the pathology in those who died was the same as in bodies which had died from asphyxia. Cartwright 1952, p. 284.
108. Hickman 1824.
109. Bichat 1977. See also Pickstone 1981.
110. Bichat 1977, pp. 82–3, 179–80.
111. Cooter 1984, p. 3.
112. Manuel provides a useful synthesis of eighteenth- and nineteenth-century research into the nervous system which I draw heavily on in this section. Manuel 1996, Chapter 5.
113. Quoted in Manuel 1996, p. 251.
114. Flourens 1824. Porter 1999, p. 538.
115. Hooper's *Medical Dictionary* of 1798 described several 'species' of asphyxia which shared the common symptom of a body without pulsation of heart or arteries. By the time Hickman was performing his experiments, the definition of asphyxia centred on the suspension of respiration which created

insensibility. See also Albury 1993. Davy's brother explained after Davy's death, that no post-mortem had been performed because of Davy's idea:

> that it was possible for sensation to remain in the animal fibre after the loss of irritability and the power of giving proof to others...he also had a horror of being buried alive, before animation was completely extinct, and he desired that the internment should not be performed till after 10 days.

John Davy, *Memoirs of the Life of Sir Humphry Davy Bart* (London: Longman, 1836) quoted in Bergman 1998, p. 267.

116. *L* II (1825–26) 647.
117. French 1975, Rupke 1987.
118. Rey 1995, p. 183.
119. French 1975, p. 20.
120. Desmond 1989, Manuel 1996.
121. Desmond 1989, Jacnya 1982.
122. Desmond 1989, p. 129.
123. *British and Foreign Medical Review*, 9 (1840) 98.
124. Cooper 1829, p. 105.
125. Stanley 2003, pp. 87–8.
126. Ibid., p. 88.
127. *L* II (1844) 412–13.
128. Liston 1838, p. 7. On French surgeons' 'economy of pain', see Rey 1995, pp. 141–3.
129. Wilson 1866, p. 151.
130. *L* I (1840–41) 25–9, *L* II (1840–41) 228–9, *L* I (1841–42) 240, 326, *L* II (1845) 367–8. Ackernecht for example, argues that 'future developments in surgery awaited on the discovery of adequate pain-control methods'. Ackernecht 1982, p. 188.
131. Quoted in Stanley 2003, p. 284.
132. Rey 1995, p. 93.
133. On, Cline see Stanley 2003, pp. 269–70, on Burney, see Hemlow 1975, Vol. 5, pp. 600–8.
134. *L* II (1833–34) 596.
135. Curling 1848, p. 24.
136. For many examples of pre-anaesthetic operations which show how fortitude was built into surgical expectations of patient behaviour, see Stanley 2003, Chapter 10.
137. *L* I (1839–40) 960–1.
138. Berridge and Edwards 1981, Chapters 3,4,5,9 and Hayter 1971, Chapter 1.
139. Quoted in Rey 1995, p. 141. A similar view was expressed by New York surgeon Valentine Mott that painless surgery was a 'chimera that we can no longer pursue in our times'. Quoted in Pernick 1985, p. 72.
140. Porter 1999, p. 12. On the nature of scientific discoveries, see Ogburn and Thomas 1922 and Schaffer 1986.
141. Hooper 1839. See also *L* II (1846) 144–5.
142. *L* I (1832–33) 705–9.
143. In this section, I draw heavily on Winter 1998, particularly Chapters 3,7,8. See also Gauld 1992.

144. Winter 1998, p. 169.
145. Laing Gordon 1897, p. 95, Thatcher 1953.
146. On innovation, see Pickstone 1992, Introduction.
147. Gauld 1992, p. 281.
148. Ibid., p. 298.
149. Braid 1843.
150. Manuel 1996, p. 266.
151. Braid 1843, p. 251.
152. Ibid., p. 252.
153. Elliotson 1846, pp. 67–8.
154. The most recent study is that by Wolfe which provides a highly detailed narrative of the 'ether controversy' yet focuses little on the wider context of the discovery. Wolfe 2001.
155. Pernick 1985, pp. 66–8. The status of dentistry was being contested in many parts of the world during this period. See *Littell's Living Age,* 18 March 1848, 544–5, Hillam 2003.
156. Duncum 1994, pp. 89–93, Wolfe 2001, pp. 500–4, Young 1942.
157. Valerius Cordus (1515–44) synthesised sulphuric ether around 1540 and named it 'sweet oil of vitriol'; it was renamed 'aether' by W. G. Frobenius in 1730. See Duncum 1994, p. 63. The similarity of the effects of nitrous oxide and ether were noted in 1818 in an article attributed to Michael Faraday of the Royal Institution. Smith 1982, p. 33.
158. De Ville notes how the Southern states of America remained 'pessimistic and stoical' during this period whilst the rest of the country grew 'increasingly optimistic and confident of the prospects of human and social perfectibility'. He suggests that this may well explain the lack of local interest in Long's experiments. De Ville 1990, pp. 134–5.
159. *Hooper's Medical Directory* 1839 described how the inhalation of gases destroyed the irritability and sensibility of nervous system. In 1846 the American *Western Medical Journal* carried a warning against the 'pernicious effects of the inhalation' of the vapour of sulphuric ether. *L* I (1846) 605.
160. The entrepreneurial side of Morton's character and his quest for financial security is stressed by Wolfe 2001.
161. Wolfe 2001, p. 45.
162. Hargreaves 2003.
163. *Littell's Living Age,* 18 March 1848, 532. Dental advertisements of the period frequently referred to the problem of pain. See for example Mr Moggeridge who practised in Bond Street, London who advertised his 'painless yet effective system' for fitting artificial teeth in the *Lancet* in January 1846. *L* I (1846) 28.
164. Wolfe 2001, p. 45.
165. Smith 1982, pp. 54–8, Wolfe 2001, pp. 50–3.
166. Wolfe 2001, p. 144.
167. A full account is given by Wolfe 2001.
168. Wolfe 2001, p. 145.
169. See Duncum 1994, pp. 100–102 for accounts of Morton's experiments.
170. See for example Morton's advertisement for ether: 'Teeth extracted without pain', underwritten by 'certificates of medical men of the highest authority,

that the inhalation is attended to with no injurious effects'. Quoted in Wolfe 2001, p. 119.

171. See Ellis 1976, 1977.
172. Quoted in Pernick 1985, p. 82.

## 2 Altered states

1. Armstrong Davison 1965, p. 117.
2. For a reading which constructs Bigelow's strong defence of Morton's attempts to patent ether as an expression of the inherent tensions of the professionalisation of American medicine and the country's democratic culture, see Browner 1999.
3. Henry Jacob Bigelow, *Paper read to Boston Society of Medical Improvement, Nov 9th 1846, an abstract having been previously read before the American Academy of Arts and Sciences, Nov 3rd 1846*, reprinted in *L* I (1847) 6–8.
4. *L* I (1847) 5. In 1830, Boott had been a contender for the chair of physics at University College London which was subsequently awarded to John Elliotson, advocate of mesmerism. Ellis 1976, 1977.
5. *L* I (1847) 8.
6. Winter 1991, 1998, Cock 1911, Poore 1872.
7. Stanley 2003, pp. 64–5.
8. Desmond 1989, p. 127.
9. Winter 1998, p. 94. In practical terms, the prolonged period often required to establish a mesmeric trance jarred with Liston's emphasis on speed. See Stanley 2003, p. 223.
10. *L* I (1847) 5, Brodie 1821, *Boston Medical and Surgical Journal*, 38 (1848) 229–37.
11. See for example Christison 1845 which classified sulphuric and nitric ether as poisons.
12. *LMG* 38 (1846) 1089.
13. *L* I (1847) 52.
14. *MT* 16 (1847) 24.
15. Quoted in Robinson 1983, p. 28.
16. Walker 1996.
17. Robinson 1983, p. 43, Walker 1996.
18. Secher 1990, Wilson 1995.
19. Robinson 1983, p. 32.
20. Ibid., p. 31.
21. Quoted in Lane 2000, p. 111.
22. Robinson 1983, pp. 21–2. See also an account of operations at St George's on 25 January 1847. After watching three operations under ether, one onlooker – Lord Morpeth – 'inhaled the vapour but did not seem inclined to go beyond tasting its flavour'. *MT* 16 (1847) 68.
23. *Boston Medical and Surgical Journal*, 35 (1846) 316.
24. Ibid. (1846) 324.
25. Winter 1998, p. 180.
26. Ibid., pp. 174–5.
27. Ibid., p. 176.

28. *British and Foreign Medical Review*, 23 (1847) 312. James Braid, pioneer of 'nervous sleep' understood ether to produce the same results as mesmerism; ethereal narcotism, he said, suspended pain 'much more generally and rapidly' than mesmerism. *MT* 16 (1847) 10 and 130.
29. Here I draw particularly on the work on science by Bynum 1994, Cannon 1978, Golinski 1992 and Pickstone 2005.
30. Vincent 1981, p. 173.
31. Cannon 1978, Chapter 1.
32. Henry Brougham, *A Discourse on the Objectives, Advantages and Pleasures of Science* (London, 1827), p. 41 quoted by Vincent 1981, p. 142. This was the first publication of the Society of Useful Knowledge which was formed by middle-class radicals in 1826 with the objective of advancing the knowledge of the working classes.
33. In this part, I draw heavily on Winter 1998, particularly Chapter 2.
34. F. Sitwell, *What is mesmerism?* (London: Bosworth and Hamson, 1862) 2nd edn. pp. 5–6, quoted in Hilton 1995, p. 232.
35. An undated letter from Elizabeth Gaskell to Ann Scott, wife of Alexander Scott, principal of Manchester's Owens' College refers to the possible use of mesmerism on an acquaintance who has been discharged from St Bartholomew's Hospital with an 'incurable' tumour at the mouth of the womb. Gaskell asks Scott if she or her husband have any experience of mesmerism because she adds: 'I have rather a dread of it altogether I *think* because I have a feeling that it *twisted* Miss Martineau's mind; but it may not be that, & it may be a superstitious feeling'. Hilton suggests that the likely date of the letter is in the early 1850s. Hilton 1995. See also discussion of Harriet Martineau and mesmerism in Cooter 1991.
36. *LMG* 4 (1847) 394–5.
37. *Littell's Living Age*, 12 June 1847, 493.
38. My figures are taken from the accounts found in Robinson and exclude tooth extractions.
39. *MT* 15 (1847) 333.
40. Winter estimates that between 1841 and 1846 around 60 operations and 200 tooth extractions were performed under mesmerism, although she notes that this may well be an underestimation. Winter 1991, p. 16.
41. See an account of an operation at St Thomas' hospital which was aborted because the ether had produced 'so much coughing, turgidity of the face, quickened circulation... threatening congestion of the brain and lungs'. Robinson 1983, p. 33.
42. Robinson 1983, pp. 25–6.
43. Ibid.
44. Ibid., p. 15.
45. On therapeutic practice, see Bynum and Nutton, 1991, Rosenberg 1979, Warner 1996.
46. Robinson 1983, p. 45.
47. Ibid., p. 9.
48. *L* I (1847) 6–8.
49. Nicolson 1993, Porter 1993, Reiser 1978.
50. Robinson 1983, p. 25.
51. *L* I (1847) 8.

52. The experiments on the O'Key sisters at University College hospital in the late 1830s using animal magnetism had been entirely discredited when it was found that the effects derived from their 'imagination'. See Winter 1998, pp. 98–9.
53. *MT* 15 (1847) 324.
54. Durey suggests that around this period, many new therapies were abandoned because of the difficulties encountered in confirming their efficaciousness: experimentation on hospital patients was very unpopular and doctors were reluctant to jeopardise their reputation by giving new therapies to private patients. Durey 1983, p. 271.
55. On the perception of a 'decline' in English medicine during the first part of the 19th century, see Warner 1991.
56. *London Medical and Surgical Journal*, 4 (1830) 89. See also: 'It is passing strange that our *Pharmacopoeia* should always be behind the progress of science', *London Medical and Physical Journal*, 8 (1832) 14, quoted in Durey 1983, p. 271.
57. *L* I (1847) 52.
58. *L* I (1847) 150, *L* I (1847) 392–3, *MTG* 15 (1847) 289–90.
59. Porter 1999, pp. 253–4.
60. *OC* 348.
61. Quoted in Duncum 1994, p. 159.
62. Ibid., p. 153.
63. On Snow's early years and medical training, see Snow 1995, 2000, Vinten Johansen 2003.
64. Van Swanenberg 1983.
65. Peterson 1978.
66. See for example: 'Mechanism of respiration', *L* I (1838–39) 653–5, 'On asphyxia and on the resuscitation of still-born children', *LMG* 29 (1841–42) 222–7, 'On the circulation in the capillary blood-vessels, and on some of its connections with pathology & therapeutics', *LMG* 31 (1842–43) 810–8, 'On the pathological effects of atmosphere vitiated by carbonic acid gas', *Edinburgh Medical and Surgical Journal*, 65 (1846) 49.
67. In 1843 he had used ether as one of several volatile agents to investigate the capillary circulation in a variety of living creatures. Snow 1995.
68. 'On paracentesis of the thorax', *LMG* 29 (1841–42) 705–7. See also *CB* 13.2.1850 for an account of his use of the trocar and cannula.
69. For Charles Darwin's response to surgery, see Browne 1995, Desmond and Moore 1992. For other contemporary descriptions, see Paget 1901, Ward Richardson 1897.
70. Many doctors supplemented their income with appointments to dispensaries and friendly societies where patients paid for medical care through a subscription. See Digby 1994, pp. 50–1 and Lane 2001, pp. 78–9. The Aldersgate School of Medicine rivalled St Bartholomew's in the 1830s but was unable to attract enough students and eventually closed in 1849. Snow taught one course of Forensic medicine during the summer session and would have earned between £40 and 60 out of which he would have needed to fund equipment and experiments. Snow 1995.
71. *MT* 15 (1847) 274.
72. Dalton 1805.
73. *LMG* 6 (1842) 5–6.

74. *LMG* 4 (1847) 156–7 and *L* I (1847) 99–100. On ineffective inhalers, see *LMG* 4 (1847) 525 and *LMG* 5 (1847) 85.
75. *LMG* 4 (1847) 501–2.
76. *L* I (1847) 99–100.
77. *L* I (1847) 120–1, 158, 184. See also comments by Plomley who watched Snow's administrations: 'the apparatus ... has many advantages; but its success may be found ... to depend more upon the great tact, and thorough knowledge of the effects of the vapour upon the system ... [of] Dr Snow, than upon any peculiarity in its arrangement and mechanical construction'. *L* I (1847) 159.
78. *LMG* 4 (1847) 384.
79. *LMG* 4 (1847) 383.
80. *LMG* 4 (1847) 514.
81. *L* I (1847) 227–8 and *LMG* 4 (1847) 383–4.
82. *L* I (1847) 228.
83. *L* I (1847) 259. He chose birds for many of his experiments as 'they breathe more air in proportion to their size than other animals, and, consequently, inhale the vapour more quickly', *L* I (1847) 551.
84. *The Inhalation of Vapour of Ether in Surgical Operations*, lecture given to medical members of the United Services Institution on 12 May 1847. Reprinted in *L* I (1847) 551–4.
85. *L* I (1847) 553.
86. *OE* 32–3.
87. Ibid. 33.
88. *L* II (1847) 410–1, *LMG* 5 (1847) 812–4. Snow wrote to the *LMG* pointing out that they had misquoted from his text on two occasions in the review. His concern was that, 'it caused you [*LMG*] to give an account of the efficacy of ether directly opposed to the facts I endeavoured to explain', 'Dr Snow on the Effects of Ether Vapour', *LMG* 5 (1847) 859. The *LMG* apologised for the first misquote but stated that the second had been in a paragraph not placed between quotation marks and was thus not intended to be a verbatim report, *LMG* 5 (1847) 898–9.
89. Robinson 1983, p. 12. Here Robinson was drawing on Francis Boott's impressions of the first trials of ether. See pp. 11–12. Until March 1847 Robinson was very active in promoting the benefits of ether and demonstrating its practice but around the time of the publication of his book, he fades from prominence. See Hillam 1996.
90. *OE* 1–2.
91. Ibid. See also *LMG* 7 (1848) 333–5 and 412–4.
92. It reflects the new surgical view of the body – post-mortems matched lesions with the external pathology of disease – but it was not until the 1870s that the work of French physiologist, Claude Bernard, established how drugs acted on specific sites within the body.
93. *OE* 2.
94. Ibid. 35–6.
95. See account by James Startin on the 'three principal stages' of etherisation: excitation; unconsciouness and insensibility; faintness and complete syncope. *MT* 16 (1847) 85, also *L* I (1847) 345–6.
96. Duncum 1994, pp. 160–1.

97. Ibid., p. 158.
98. Ibid., pp. 158, 170.
99. An account of the inquest was carried in the *Lancet*. *L* I (1847) 340–2. See also *Times*, 19 March 1847, page 8, column f and 30 March 1847, page 6, column a.
100. *L* I (1847) 343.
101. *MT* 16 (1847) 101.
102. *L* I (1847) 349.
103. On the difficulties experienced by Welsh doctors, see Owen 1994, Lewis 1996, on the decline in Bristol, see Bennett 1999, on Aberdeen, see Keith 1849.
104. Keith 1849.
105. *OE* 5, 33 and *OC* 19.
106. Simpson 1847.
107. Letter from Duncan to Sir Robert Christison, 6 March 1875, quoted in Wilkinson 2003.
108. Waldie had suggested the use of chloroform as an anaesthetic to Simpson in October 1847, but owing to various difficulties following the destruction of his laboratory by fire, had not followed up the conversation with a sample. As a result, Simpson procured some from Edinburgh chemists, Duncan and Flockhart. Those with a detailed knowledge of chloroform's history, such as Snow, were able to distinguish the importance of Waldie's contribution to the discovery as his suggestion was based on 'an almost certain knowledge of its [chloroform's] effects'. But after Simpson's death in 1870, both Waldie and Duncan made public their sense of unhappiness with the lack of recognition they had been accorded at the time. See *OC* 17, 29, Waldie 1870, Wilkinson 2003.
109. On Lawrence's use of chloric ether, see *LMG* 5 (1847) 939, 1153, *L* II (1847) 571. There were still disputes about the correct name of chloroform into the 1850s when the London *Pharmacopoeia* listed it as chloroformyl, a name said Snow, 'which departs from the brevity of the word chloroform without having the merit of expressing the chemical constitution of the substance'. See *OC* 29.
110. Warren 1849, p. 5.
111. *LMG* 5 (1847) 1030–31 and *L* II (1847) 575–6.
112. *MT* 16 (1847) 24.
113. *LMG* 5 (1847) 1030–31 and *L* II (1847) 575–6.
114. *LMG* 6 (1848) 255. A report of the inquest appeared in *LMG* 6 (1848) 250–5, 283–4. On Greener's death, see Knight and Bacon 2002.
115. Duncum 1994, pp. 195–7. On Glover's work, see Defalque and Wright 2004.
116. *LMG* 6 (1848) 253.
117. *L* I (1848) 175–6.
118. *LMG* 6 (1848) 277–8.
119. *LMG* 7 (1848) 108. See also Sibson's article: 'Remarks on the action of narcotic poisons' in *LMG* 6 (1848) 266–71. He developed a facepiece to give ether which Snow later adapted. See *LMG* 4 (1847) 363 and *OE* 22.
120. *LMG* 6 (1848) 277–8.
121. *OC* 120–36.

122. *OC* xiv. He explained his rationale as 'an occasional risk never stands in the way of ready applicability'.
123. *LMG* 7 (1848) 841. See also, 'On the administration of chloroform in the public hospitals', *MTG* 4 (1852) 349–50, 'Death from chloroform', *BMJ* I (1858) 279.
124. *LMG* 7 (1848) 842.
125. Bland Sutton 1930, p. 69.
126. Clendon 1849, pp. 10–11.
127. Duncum 1994, p. 11.
128. On 'heroic' nature of American practice, see Pernick 1985, Warner 1996.
129. Quoted in Duncum 1994, pp. 204–7.
130. Here I would differ from Winter who constructs the difference between mesmerism and ether to be one of medical control: 'ether anaesthetists were able to secure tacit agreement from their audience that they controlled the experimental situation. Pain . . . had become the property of the surgeon, to be "conquered" by the anaesthetic apparatus at will', 1998, p. 179. Though this reading may well apply to autocratic individuals such as Liston, for the majority of doctors the patient's cooperation with the process of inhalation was essential.
131. Robinson 1983, p. 32.
132. *MT* 15 (1847) 311.
133. Ibid. 16 (1847) 122.
134. *L* I (1847) 227–8.
135. Ibid.
136. Quoted in Duncum 1994, p. 160.
137. Ibid. pp. 160–1.
138. *LMG* 7 (1848) 108.
139. Ibid.
140. Duncum 1994, pp. 184–5.
141. *OC* 120–1.
142. Bickersteth 1853, Black 1855.
143. Robinson 1983, pp. 39–40.
144. Ibid., p. 51. See also: 'contortions of the face and limbs . . . do not physiologically prove pain; they may take place . . . from mere reflex-action'. Brookes 1847, p. 19.
145. Quoted in Robinson 1983, pp. 7–8.
146. *Littell's Living Age*, 12 June 1847, p. 486.
147. Ibid.
148. Fissell 1992.
149. Robinson 1983, pp. 33–4.
150. *OE* 51–2.

## 3 Science versus empiricism

1. Fergusson 1852, p. 24.
2. On meanings of nineteenth-century science, see Bynum 1994, Cannon 1978, Pickstone 2005.

3. One notable study on the emergence of 'pain medicine' in France during the second half of the twentieth century shows how two groups of actors shared a common model and theory of pain medicine but applied it in different ways. It takes the differentiation around a set of common principles as confirmation of the diversity and dynamism of medicine. See Bazanger 1998.
4. On Snow, see Snow 1995, 2000, Vinten-Johansen 2003; on Simpson, see Shepherd 1969, Simpson 1972.
5. On Snow's dispute with Lonsdale, see *LMG* 23 (1838) 415–7, 559–60, 640–1 and 718–20. For an account of a disagreement between Lee and the Royal Society on the nerve supply to the uterus which lasted for around 10 years, see Manuel 2001.
6. See for example *LMG* 1 (1845) 1009–06.
7. *L* II (1847) 575–6.
8. Ibid. (1855) 560.
9. *MT* 15 (1847) 274, Laing Gordon 1897, pp. 102–103.
10. Simpson 1972, p. 110.
11. *OE* vi. Simpson did experiment but usually upon himself or others rather than using animals. For example, in 1848 he carried out some experiments on local anaesthesia by suspending his hand over a jug of chloroform, *L* I (1854) 450.
12. For details of the publishing process and the dates of each article, see Introduction in Ellis 1991.
13. *LMG* 10 (1850) 622.
14. Ibid. 6 (1848) 853.
15. Ibid. 6 (1848) 853–4 and *OC* 73.
16. Ibid. 6 (1848) 893.
17. Ibid. 6 (1848) 1074–5. See also his account of learning of bisulphuret of carbon through the medical journals, ibid. 1076–7.
18. Herschel 1830.
19. Pickstone 2000, pp. 148–9, 153.
20. Herschel 1830, p. 15.
21. Ibid., pp. 13–14.
22. See for example his early work on the use of arsenic as a preservative for dead bodies, *L* I (1838–39) 653–5.
23. Herschel 1830, pp. 174–5. For an example of Snow's application of this approach in other areas of practice, see his comments during a discussion of the endosmotic action of medicines at the Medical Society of London in 1852: 'In order to fully understand the action of medicines, other laws must be considered as well as those governing endosmosis'. *MTG* 4 (1852) 23–4.
24. Herschel 1830, p. 93.
25. See *OE* vi, 1 and Herschel 1830, p. 21.
26. On Herschel's influence on Darwin, see Browne 1995, Ruse 1975.
27. He noted wryly that

> a medical author of great reputation in Paris sent to enquire at what temperature I used the water bath [with the inhaler] and being informed, at the ordinary temperature, published his opinion that it had no effect, and might as well be left off. He appeared not to have considered the relations of heat, either to liquids or vapours.

*OC* 84. See also a dispute with Simpson over the boiling point of chloride of amyle. *MTG* 14 (1857) 457.

28. Snow 1995.
29. *LMG* 7 (1848) 842. The most accurate means of giving vapour, for example, was with a balloon but he conceded that few would be able to cope with this in clinical practice.
30. *OE* 14.
31. *OE* 34. See his comments on experiments of other doctors who asserted that in death from chloroform, the heart's action survives respiration: 'The greater number of experimenters who have killed animals with chloroform have found that the action of the heart continued after the breathing ceased; but they did not either control or ascertain the proportion which the vapours of chloroform bore to the inspired air.' He cites the experiments of Thomas Wakley and those of the French 1855 Commission, *OC* 122.
32. *LMG* 7 (1848) 841.
33. *OE* 15–23, *LMG* 7 (1848) 1022.
34. *LMG* 7 (1848) 841.
35. Ibid. 4 (1847) 395.
36. Ibid. 5 (1847) 349.
37. Simpson's method was adopted on the battlefields of the Crimea because of its ease. See Connor 1998, Shepherd 1991.
38. Simpson 1847, p. 27.
39. Ibid., pp. 29–30.
40. Miller 1848, pp. 15–16.
41. *L* II (1847) 550.
42. Miller 1848, p. 17.
43. *MTG* 4 (1852) 253.
44. *OC* 79.
45. *MTG* 4 (1852) 253–4.
46. Ibid. 598–9.
47. Ibid. 627–8.
48. Syme dismissed one death which had occurred in his hospital on the grounds that it was under the responsibility of a different surgeon.
49. The issue of the purity of ether and chloroform samples was frequently raised, but by 1850s, Scotland was established as main supplier of chloroform.
50. *L* I (1855) 55–6. See discussion of Scottish versus English anaesthetic mortality in Sykes 1960, Vol. 1, pp. 137–67.
51. *L* I (1855) 108–9.
52. For example in the 1810s, Bristol surgeon James Bedingfield noted in relation to the bleeding of patients: 'No regard was paid to the quantity taken; an abatement in the violence of the phrenetic symptoms formed the criterion by which the flow of sanguineous fluid was regulated.' Quoted in Fissell 1992 p. 105. On therapeutic practice in America, see Rosenberg 1985, Warner 1986.
53. See also: 'no patient is proof against good chloroform well administered. Some require a much larger dose than others; but, sooner or later, all will succumb', Miller 1848, p. 724
54. Miller 1848, p. 20.
55. Ibid., p. 21.

56. Ibid.
57. *L* I (1855) 108–109.
58. Ibid. 192.
59. On the contest over orthodoxy in mid-nineteenth-century medicine see Loudon 1986.
60. *L* I (1855) 192. In regard to American orthopaedic surgery and the surge in medical litigation from the 1830s onwards, see De Ville 1990.
61. *L* I (1855) 192.
62. Bird 1853, p. 17.
63. See Digby 1994, Peterson 1978, Waddington 1984.
64. *L* I (1855) 192.
65. Porter 1999, pp. 635–6.
66. Fissell 1992.
67. Paris was one of the few centres where Louis' therapeutic nihilism was more than a theory; American physicians found it impossible to put into practice and so too it would appear did London doctors. See Warner 1996.
68. *L* I (1846) 229.
69. Soon after the introduction of ether Snow wrote: 'it will be at once admitted that the medical practitioner ought to be acquainted with the strength of the various compounds which he applies as remedial agents, and that he ought, if possible, to be able to regulate their potency', *LMG* 4 (1847) 498.
70. Snow was swift to note that his inclusive approach to anaesthesia should not be judged as rash or indiscriminate; it was distinct therefore from Simpson's universality. Nor was it unconnected to clinical practice: 'supposed objections have one by one vanished', he explained, 'and it has appeared that care in the mode of giving the vapour was the main guarantee, both of safety and success', *LMG* 7 (1848) 1021.
71. In this section, I draw heavily on the work of Baker 1993 and McCullough 1993.
72. Quoted in Waserman 1980, p. 159.
73. Laing Gordon 1897, pp. 48–9. See also 'that great principle of emotion which both impels us to feel sympathy at the sight of suffering in any fellow-creature, and at the same time imports to us delight and gratification in the exercise of any power by which we can mitigate and alleviate that suffering', *L* II (1847) 550.
74. John Gregory, *Lectures on the Duties and Qualifications of a Physician*, (W. Strahan and T. Cadell: London, 1772) pp. 19–21 quoted in McCullough 1993.
75. McCullough 1993.
76. *OC* 206–8, *MTG* 5 (1852) 361–2.
77. Duncum 1994, pp. 192–3.

## 4 Risks of life and birth

1. Pernick 1985, p. 22 and particularly Chapter 5.
2. Burney 2006a.
3. Caton 1999, Poovey 1987, Youngson 1979.
4. But, see Poovey 1997.

5. Pernick 1985, p. 101.
6. Trohler 2000.
7. Ibid., pp. 16–17.
8. Variolation – the inoculation of matter from smallpox pustules into the skin – was practised from the 1720s but from the 1790s Jenner proposed that this should be replaced by vaccinations of cowpox. It proved to be a highly contentious issue in which the notion of risk played a primary part. Baxby 2004.
9. Simpson 1846, p. 6.
10. Ibid., p. 9.
11. *L* I (1847) 551–4.
12. Simpson 1871, p. 32
13. Thomas Inman, 'Tables of the Mortality after Operations', *L* II (1844) 39.
14. Simpson 1871, pp. 72–8. Snow provided Simpson with figures for amputations carried out at St George's which were all secondary cases; these showed a mortality rate of 10 per cent which was considerably lower than the 20 per cent rate across all hospitals.
15. *L* I (1855) 292. During a debate on merits of bloodletting, Crisp commented that although facts suggested that bloodletting was not a beneficial therapy, in practice patients could not be cured without it. *MTG* 4 (1852) 174–5. See also a debate on the innovation of ovariotomy and arguments against the validity of statistics. *L* I (1854) 420–1.
16. *L* I (1853) 523–4.
17. *OE* 7. Not until the twentieth century did it become a common view that anaesthetic agents metabolised within the body rather than being breathed out.
18. *L* I (1848) 166. Fergusson had become the premier surgeon after Liston's unexpected death in December 1847.
19. *LMG* 5 (1847) 812–14.
20. Druitt 1852, p. 523.
21. *BMJ* (1858) 207–8.
22. Berridge and Edwards 1987, pp. 76–84.
23. Weatherall 1993, p. 919.
24. Murphy 1848, p. 37. See also letter from John Robertson, Manchester surgeon, comparing chloroform to opium. *MT* 2 (1853) 466. Also: 'there are persons, in all respects sound and healthy upon whom chloroform, absorbed in certain quantities into the circulating fluid, acts as a poison and may cause the rapid extinction of life'. *MTG* (1852) 346.
25. Holloway 1995, p. 90.
26. For an analysis of the Victorian fear of poisoning and the emergence of the poison hunter – the toxicologist – see Burney 2006b.
27. Holloway 1995, p. 90.
28. *LMG* 12 (1851) 498–9.
29. *OC* 120–200.
30. *L* I (1853) 523–4.
31. *L* II (1854) 404–5.
32. Ibid.
33. Ibid. 513.
34. *OC* 143.
35. *Edinburgh Medical Journal* XI (1866) 645–8.

36. Longford 1964, p. 219.
37. *CB* 11.11.48, 16.4.49.
38. Holmes 1898, pp. 174–5.
39. *OC* 151.
40. See discussion at a meeting of the South London Medical Society in April 1847: Bransby Cooper, surgeon at Guy's, described pain as a 'necessary adjunct' to surgery; Nunn, the Colchester surgeon who had suffered an ether fatality, spoke of it as a 'safeguard' which protected vitality; and for Pickford, intervening to prevent pain in surgery posed a direct 'danger' to life. *LMG* (1847) 30 April 1847.
41. On individual susceptibility to pain, see Curling 1848, p. 9.
42. *MT* 16 (1847) 572.
43. Ibid.
44. Percival 1985, p. 11.
45. A patient at Charing Cross hospital: 'was so alarmed at the anticipation of an operation, that he became extremely ill and desponding, and died in a week in a state of coma'. *L* II (1848) 312. See also: 'no surgeon would, if he could avoid it, perform an operation on a person strongly prepossessed with the idea that he should never recover from it'. Curling 1848, p. 20.
46. *MT* 16 (1847) 572. See also: 'the emotion of fear supplies a great part of the pain suffered under the operating knife'. *L* I (1847) 74.
47. Few surgeons publicly acknowledged that fear of anaesthesia was a problem for some patients. Fergusson writing in 1853 stated: 'a patient can contemplate such means of cure with a calmness and seeming fortitude which even the bravest or most stoical could not maintain in former times'. Fergusson 1853, p. 22.
48. Robinson 1983 p. 7.
49. Tracy 1847, pp. 11–12, 25.
50. Murphy 1848, p. 17.
51. *L* II (1853) 411.
52. *MTG* 8 (1854) 606. *CB* 7.6.52. See also an account of a death at St George's which notes how the patient was 'overcome by fear'. *L* I (1854).
53. *MT* 2 (1853) 120.
54. *LMG* 7 (1848) 108.
55. *OC* 76–7. Snow also cites cases of patients dying before chloroform had been absorbed into their nervous system; a fact which he verified at post-mortem.
56. *LMG* 10 (1850) 327.
57. *OC* 77–8 and *CB* 5.3.55.
58. *OC* 38.
59. *LMG* 4 (1847) 514.
60. *OC* 211–1.
61. *Household Words*, VII (1853) 181.
62. *L* I (1848) 163.
63. See for example *Times* 5 October 1849, 'Chloroform, use of, by thieves', page 3, column f.
64. Berridge and Edwards 1987, pp. 81–2.
65. *Times* 15 January 1850, page 7, column c, 25 January 1850, page 7, column d.
66. *Times* 1 January 1850, page 7 column d and Snow 1851, pp. 9–11.

67. *LMG* 10 (1850) 327. Snow's letter was also reprinted in the *Times* 22 February 1850, page 5, column c.
68. *Times* 1 May 1850, page 7, column e, 23 May 1850, page 8, column 6, Snow 1851 p. 12.
69. *LMG* 11 (1850) 834–5, Snow 1851 pp. 12–13.
70. *L* I (1850) 607.
71. *LMG* 12 (1851) 498–99.
72. Ibid. 499–501 and *Times*, 15 March 1851, page 2, column c.
73. Snow 1851.
74. *LMG* 12 (1851) 571.
75. Snow 1851, p. 4.
76. *LMG* 12 (1851) 498–501. In his essay *On Liberty*, written between 1854–58, Mill noted that the issue of poisons had raised a new question; how far may individual liberty be legitimately invaded for the prevention of crime or accident? Mill 1991, pp. 106–107.
77. *L* I (1853) 505. See also: 'There is scarcely a poor hospital out-patient who does not throw out a broad hint for chloroform on the smallest suspicion of the slightest operation', *L* II (1857) 204.
78. Conan Doyle 1981, p. 953.
79. This view is substantiated by the growth in the life assurance business: in 1800 there were six life offices, by the 1850s there were 190. See Dupree 1997.
80. *Nineteenth Annual Report of the Registrar General*, 1858, pp. 197–8.
81. Simpson compared the risks of anaesthesia to those of other innovations – railways, steamboats, stagecoaches – which were all attended by accidents to 'limb and life'; he argued that this was no reason for their abandonment. 1871, p. 187.
82. *OC* 372–410.
83. *CB* 7.4.57.
84. *CB* 22.1.57.
85. *OC* 416.
86. Ibid.
87. Not all French surgeons followed the Académie's recommendations. Monsieur Giraldes, member of the Paris Sociéte de Chirurgie who was present in the operating theatre of St George's at the time of Snow's second fatality, continued to use amylene in his practice; he took the view that amylene was safer and more effective than chloroform. *OC* 416–7.
88. Richardson 1897, p. 284 and 1896, pp. 69–70.
89. *MTG* 14 (1857) 84. See also *MTG* 14 (1857) 332–4, 357–9, 379–82.
90. *BMJ* II (1857) 654.
91. *OC* 418.
92. Richardson 1897, p. 284.
93. Loudon 1992, pp. 172–3.
94. Ibid., p. 340.
95. Ibid., Chapter 7.
96. Ibid., p. 14.
97. Leavitt 1996, Chapter 1.
98. Loudon 1992, p. 343. An alternative view is given by Caton who suggests that opiates only became part of 'normal labour management' after the introduction of twilight sleep in 1902. Caton 1995, p. 780.

99. Loudon 1992, p. 173.
100. Bedford 1995, particularly pp. 171–88. Their opposition also revealed similar concerns to those voiced on the introduction of the speculum. Moscucci 1990, pp. 112–18.
101. *L* I (1847) 377. See also discussion of Smith's views in Poovey 1987, pp. 143–5.
102. Gream 1848, p. 7.
103. Simpson 1849, p. 246. Loudon notes that between the 1880s and the 1930s the view that 'civilized women' were no longer capable of tolerating the pain and stress of labour sustained a marked increase in the use of pain-relief and assisted delivery. Loudon 1992, p. 340.
104. *L* II (1841–42) 404.
105. Quoted in Digby pp. 270–1.
106. Simpson 1848. On religious opposition to anaesthesia, see Caton 1999, Farr 1983, Youngson 1979.
107. Quoted in Merriman 1848, p. 11.
108. *L* II (1847) 572–4.
109. *LMG* 5 (1847) 1031–2. See also Murphy's report of a case in February 1847 when the use of ether in labour created 'a half-drunken stupid state' in the woman. *L* I (1847) 228.
110. *LMG* 5 (1847) 1031–2. The difference between ether and chloroform was also noted in surgery. Hancock reports that after operations on strangulated hernias 'patients awoke calm and collected' from chloroform; ether caused patients to 'awake in excited state'. *LMG* 5 (1847) 1031–2.
111. Murphy 1850, p. 15.
112. Ibid.
113. *OC* 321–2.
114. *London Journal of Medicine* I (1854) 976.
115. *OC* 322.
116. Ibid.
117. Murphy 1850, p. 27.
118. See also Robert Barnes who suggested that the mother's cries took the strain off her perinaeum. L II (1850) 39.
119. Merriman 1848, p. 24.
120. *L* I (1848) 476–8.
121. *OC* 324.
122. Ibid. 320.
123. *AMJ* (1853) 500–2. On nineteenth-century views on placental transmission, see Caton 1999, pp. 71–86.
124. Gream 1848. Tyler Smith asked Snow for advice on using amylene in childbirth. See *OC* 395 and *CB* 15.7.57.
125. Snow recommended the employment of a second doctor to give chloroform although he acknowledged it was quite possible for an accoucheur to manage single-handed. *OC* 323.
126. Ackland, Wellcome MS 7207.
127. *OC* 328.
128. Donnison 1977, pp. 21–2, 35, 37, 42 and Loudon 1992, pp. 177–8.
129. *CB* 28.12.48.
130. Ibid. 16.11.49.
131. Ibid. 3.2.51.

132. Ibid. 11.12.48.
133. Digby 1994, p. 255.
134. *CB* 18.6.50.
135. RA VIC/Add A7/7.
136. Snow 1995.
137. *AMJ* (1853) 318.
138. *MTG* 6 (1853) 526–7.
139. *L* I (1853) 453. See also *AMJ*'s response to the *Lancet's* criticism of the use of chloroform. *AMJ* (1853) 450.
140. *AMJ* (1853) 575.
141. *AMJ* (1853) 500–2.
142. *L* I (1857) 410.
143. *L* II (1858) 555–6.

## 5 Anaesthesia in London: John Snow's casebooks

1. Greene 1979, p. 10. See also Ackernecht 1953, pp. 5–12, Trohler 1993.
2. Snow's original casebooks are held by the Royal College of Physicians, London. Richard Ellis published a transcription in 1994.
3. On using hospital records as historical sources, see Risse and Warner 1992.
4. Different handwriting in surgical records indicates a change in the record-keeper, and in many instances, this also corresponds to a qualitative difference in the data.
5. Some writers have assumed that no reference to chloroform in surgical records meant that ether was used instead. See for example, Chaloner 2001. I have found no evidence which suggests this was the case.
6. On hospitals, see Abel-Smith 1964, Woodward 1974.
7. Knight 1851, p. 339.
8. Cameron 1954, p. 49.
9. Peterson 1978, p. 139.
10. Ibid., p. 174.
11. Ibid., pp. 149–50.
12. *L* II (1846) 335–48. Digby notes there were 951 medical students in London in 1841. 1994, p. 13.
13. Paget 1901, p. 155.
14. One of the key factors behind the growth in middle-class patients was the revolution in travel brought by the railways. They brought a 'profound' transformation to medical practice by enabling doctors to easily extend their practice beyond the suburbs of London: it was also a lucrative exercise – doctors usually charged 2/3 mileage as travelling expenses. Loudon 2001.
15. The ratio for surgical patients in Guy's during 1845 was 73 per cent males to 27 per cent females; in King's College during 1846 the ratio was 57 per cent males to 43 per cent females. *L* II (1846) 360. There is no evidence, however, that London experienced the same dramatic rise in number of accident victims as New York and Philadelphia did. See Pernick 1985, p. 219.
16. Peterson 1978, p. 175.
17. *Household Words* 2 (1851) 460.

18. Paget 1901, p. 157.
19. Liverpool surgeon John Halton noted that whilst 1480 cases of severe accidents and fractured bone had been admitted to the Liverpool Infirmary in the 1830s, only 220 major operations had been performed. Halton 1843.
20. 'Report of the Clinical Society from January 1845 to March 1846', *Guy's Hospital Reports*, Vol. 11, 4 (1846) 96.
21. Between 12 November 1843 and 30 November 1836 Liston performed 16 operations; between 17 May 1839 and 1 March 1840 he performed 39 operations. Merrington 1976, p. 28, Stanley 2003, p. 91.
22. KCH/CN/68.
23. 'Half-yearly Reports of the Clinical Society for 1844', *Guy's Hospital Reports*, Vol. 10, 3 (1845) 256, KCH/CN/68.
24. Figures published by the physician Forbes Winslow on the overall mortality in the hospitals of Guy's and St Thomas' in 1839 showed a notable drop since the beginning of the nineteenth century: in 1813 there was a ratio of 1 death to every 16 patients, by 1837 this had dropped to 1 death in 48. Quoted in Stanley 2003, pp. 152–3.
25. Simpson 1846, p. 6.
26. 'Report of the Clinical Society from January 1845 to March 1846', *Guy's Hospital Reports*, Vol. 11, 4 (1846) 96.
27. Stanley 2003, pp. 153.
28. Halton 1843.
29. Quoted in Stanley 2003, p. 192.
30. Stanley 2003, p. 193.
31. *L* II (1846) 521.
32. *L* I (1846) 592.
33. *L* I (1846) 313–14.
34. The patient eventually claimed a miracle cure had been achieved through the use of Holloway's ointment and although there may well be something of a 'tall story' about this example it does suggest that some patients found it very easy to access hospital treatment. *L* I (1846) 305.
35. Quoted in Stanley 2003, p. 194.
36. 'Report of the Clinical Society from April 1846 to March 1847', *Guy's Hospital Reports*, Vol. 12, 5 (1847) 156.
37. *L* I (1846) 393–4.
38. William Fergusson, *Daybook, 1847*, held by the Royal College of Surgeons of England.
39. *L* I (1846) 363.
40. Paget had been an assistant surgeon for a month before he undertook his first public operation at St Bartholomew's in March 1847. Although it was 'rather a difficult case', [I] 'achieved it without disgrace', he told his brother; he believed it earned him his first private operation. Paget 1901, pp. 155–8.
41. Robinson 1983, pp. 34–5.
42. Under ether, the operation took 6 minutes compared to his previous case which had lasted for 26 minutes. Robinson 1983, pp. 34–5.
43. On the advantages of anaesthesia for reducing dislocations, see Cooper 1851, p. 362.
44. *OE* 54.

45. Robinson 1983, p. 33, *L* I (1847) 79. See also Snow: 'It keeps patients still, who otherwise would not be', *OE* 53.
46. *OE* 28.
47. Liston 1846, p. 191.
48. *OE* 34
49. KCH/CN/69.
50. St Bartholomew's, St George's, King's College, University College, Guy's, the London and the Middlesex hospitals all began to use it in weekly operating sessions.
51. *L* I (1847) 211.
52. Robinson 1983, p. 31.
53. See report of patient attending Charing Cross and asking to use ether whilst his toenail was removed. *L* I (1847) 105.
54. *OE* 56–71.
55. *L* I (1847) 392. See also Tyler Smith who spoke in March 1847 of a 'general rush towards the operating room', quoted in Pernick 1985, p. 209.
56. KCH/CN/69.
57. *OE* 68.
58. KCH/CN/69.
59. On the difficulties of minimising extraneous factors which may have influenced any shift in surgical practice, see Pernick 1985, pp. 210–11.
60. *OC* 291–3.
61. Quoted in Stanley 2003, p. 219.
62. Pernick 1985, pp. 210–11.
63. G/S1/22–30, 4 September 1850.
64. *OC* 152.
65. The Hospital Minute Books at St George's record a variety of administrative decisions concerning ether and chloroform such as allocating the duty to the Assistant Apothecary, and electing Snow as an Honorary Governor in respect of his anaesthetic services to the hospital. *St George's Hospital Minute Book*, 8 July 1846–15 October 1850, pp. 629, 658–9.
66. Lloyd did use chloroform at other times, see *MTG* 5 (1852) 348.
67. MR16/10/366, St Bartholomew's.
68. Fergusson 1852, p. 22.
69. *CB* 6.1.55.
70. *OC* 76–7.
71. LH/M/3/74, London Hospital.
72. De Ville 1990.
73. Arnott 1854a, 1854b, Quinton 1856.
74. Quoted in Connor 1998, p. 163.
75. Comment by Dr Mouat, Deputy Inspector General, at a meeting of the Military Medical and Surgical Society, *L* I (1856) 79.
76. *Times* 13 October 1854, page 5, column e.
77. Ibid. 12 October 1854, page 9, column a.
78. Curling 1848, pp. 23, 33.
79. LH/M/3/74.
80. Quoted in Connor 1998, p. 178.
81. Ibid., p. 187. A similar shift is also apparent in the views of the surgeon George Guthrie, who had gained his experience of wartime surgery in the

battles of the Peninsula in the 1800s. Whereas he initially expressed caution on the use of chloroform in the 1853 edition of his work on wartime surgery, the 1855 edition included an addendum expressing his opinion that it was advantageous in all amputations. Guthrie 1853, 1855.

82. Connor 1998, p. 191.
83. *L* I (1865) 652.
84. Snow wrote to Professor Brande, in December 1852: 'I shall have pleasure in calling on you at two o'clock on Friday next in order to administer chloroform and prevent you from having the pain of the operation'. Letter to Professor Brande, 22 December 1852, St George's Hospital Medical School Archives. He duly recorded that the operation to remove some decaying bone from the tibia had gone well, and 'Mr Brande awoke and thought that the operation had not been commenced'. *CB* 24.12.52.
85. In 1848, Snow gave chloroform to Mr Birkbeck at Norfolk whilst the surgeon, Aston Key, raised portions of his skull. Birkbeck's own doctor was in attendance and 'objected to the chloroform, on account of the feeble and irregular action of the heart'. Nevertheless, his view was overridden in favour of the patient's. *CB* 21.12.48 and *OC* 301.
86. *CB* 8.6.53.
87. *OC* 259.
88. Granshaw 1992, p. 239.
89. See for example, *CB* 3.10.55 and 3.2.57.
90. *CB* 22.4.54.
91. During 1850, St Bartholomew's surgeon Stanley performed several face and mouth operations in hospital practice without chloroform. Although it was difficult in many instances to accommodate the face-piece and inhaler, it is notable that only a few months earlier, he had removed a cyst from the facial nerve in a surgeon from Salisbury, while Snow administered chloroform. *CB* 3.10.55 and 3.2.57.
92. Barker 1994, p. 934.
93. Davis 1980, Lufkin 1938.
94. *CB* 27.1.49.
95. *OC* xlii–xliv.
96. See for example, *Times* 8 September 1858, page 5, column f.
97. *L* II (1858) 407.
98. Ibid. 457.
99. See for example Henry Potter, 'Cautions in the administration of chloroform', *L* II (1858) 32–4 and letter *L* II (1858) 289.
100. There is evidence to show that some provincial hospitals also experienced a dramatic rise in the numbers of operations performed post-anaesthesia. At the Royal Berkshire Hospital, for example, whereas 17 operations were performed during 1846, 73 operations were performed in 1860. See Galland 2003, 394.
101. In Snow's words, surgical pain-relief gave surgeons the opportunity to 'explore and remove diseased joints and portions of diseased bone by operations that would be too long and too painful to be endured in the waking state'. *OC* 277–8.
102. *L* II (1865) 174.

103. Snow was under no illusion that the purpose of such plastic operations was to mitigate deformity and directly connected them to the availability of anaesthesia: without insensibility, he said, the pain would have prevented the majority of patients tolerating such procedures. *OC* 300.
104. In part, confidence was bolstered by the fact that none of the recorded chloroform fatalities had been young children. By 1857 Snow had anaesthetised 186 infants less than a year old, and 13 per cent of his patients were children of 10 years and under. *OC* 49.
105. *L* II (1841–42) 259.
106. At King's College during 1845–46, only 8 per cent of females had received operations, compared to 15 per cent of male patients. By 1859, this ratio had altered significantly. Among the female patients, 76 per cent now received operations compared to 45 per cent of male patients. KH/CN/68, 84, 85. At Guy's, hospital reports show that although numbers of surgical female admissions increased by 2.9 times between 1846 and the 1860s, operations on females increased more – by 3.7 times. During the same period, the ratio between male admissions and male operations remained very similar – admissions rose by 2.2 times, operations by 2.1 times – and a greater proportion of the 'major' operations continued to be performed on males, reflecting the male bias towards serious injury from accidents. *Guy's Hospital Reports*, 12 (1847) 156 and 15 (1870) 600–700.

## 6 In the name of safety

1. *L* II (1897) 1376.
2. On inquests and anaesthetic deaths, see Burney 2000, pp. 137–64.
3. Mason Warren 1867, p. 618. Similar arguments were made by J. E. Petrequin, surgeon in Lyons, who also abandoned chloroform in 1849 'I abandoned the use of so dangerous an agent, which gave no security to the operator and constantly exposed patients to the risk of a death always unpredictable and almost always irrevocable, and which must sooner or later occasion the operator remorse', quoted in Duncum 1994, p. 206.
4. Mason Warren 1867, p. 621.
5. Duncum 1994, pp. 352–5.
6. Duncum 1994, p. 535.
7. *L* II (1873) 879–80.
8. 'Abstract of the Report of the Committee on Chloroform', *L* I (1864) 69–72 and see discussion in Duncum 1994, Chapter 9.
9. Duncum 1994, pp. 237–40.
10. Ibid., pp. 241–2.
11. There are occasional examples of individuals who combined ether and chloroform prior to the 1860s. See for example, Bristol surgeon, Lansdown, who described using 1 part chloroform to 2 of ether for obstetric anaesthesia. *L* II (1848) 309–10.
12. *Medical Chirugical Transactions* 47 (1864) 339–43. See also *L* I (1864) 69–72. Snow had dismissed the possibility of combining the agents in the 1840s because of the difficulties created by the different volatilities; ether was the

first of the two to evaporate, thus inducing the same excitability in the patient as ether alone did.

13. Bryant 1872, p. 1023.
14. *L* I (1866) 288.
15. Duncum 1994, pp. 258–64.
16. Ibid., p. 255.
17. See Moscucci 1990, pp. 126–7. One of most famous cases concerned William Wilde, father of Oscar Wilde, who was accused of raping a patient under chloroform in the late 1850s. Ellmann 1987, pp. 13–14. See also Buxton 1888, pp. 149–51.
18. De Ville 1990, p. 31. He also cites a case in the 1880s in which the doctor was sued for using chloroform rather than cocaine as an anaesthetic for eye surgery. p. 321.
19. *OC* 200.
20. Duncum 1994, pp. 232–3.
21. Richardson 1897, p. 305.
22. Smith 1982, p. 93.
23. Ibid., 1858.
24. On the revival of nitrous oxide, see Duncum 1994 and Smith 1982.
25. Duncum 1994, pp. 273–9. An 1863 report noted that fever had been treated in America by the inhalation of the gas, *L* I (1863) 257 and a report of 4000 dentistry cases using nitrous oxide in Philadelphia appeared in 1864, *L* I (1864) 116.
26. Evans was based in Paris and had met Colton whilst the latter was visiting an international exhibition in the city. Colton taught Evans his method of making nitrous oxide. Duncum 1994, p. 279.
27. *British Journal of Dental Science* 11 (1868) 318–20.
28. Coleman 1862, p. 12 and 1881, p. 255.
29. Plomer 1986, p. 112.
30. *L* I (1868) 507–8.
31. Ibid.
32. Evans' donation was made expressly so that: 'the largest possible number, and especially the poor of London, should benefit by the advantages'. *L* II (1868) 780.
33. *L* II (1871) 687.
34. Duncum 1994, p. 285.
35. Ibid., pp. 356–63.
36. In 1868, physiologists, John Burdon Sanderson and John Murray, compared nitrous oxide to pure nitrogen by giving the gases to a small number of dental patients at the Middlesex Hospital; Sanderson also performed similar comparisons using animals. They proved that nitrous oxide anaesthesia was produced by a different mechanism to asphyxia and thus the asphyxial symptoms were a side-effect. Nevertheless most doctors, including the French physiologist Claude Bernard, remained convinced that nitrous oxide anaesthesia was produced through the mechanism of asphyxia. Duncum 1994, p. 284. Henry Davis' anaesthetic guide of 1887 was amongst several to note that 'we are not fully acquainted with its modus operandi as various opinions are held by competent investigators'. Davis 1887, p. 30.

37. *L* I (1870) 449–50.
38. Duncum 1994, pp. 264.
39. Ibid., pp. 264–5. Writing of it at the turn of the century, Frederic Hewitt commented that: 'the anaesthesia produced by "methylene" is, in fact, comparatively superficial, and would hardly satisfy most surgeons of the present day', Hewitt 1901, p. 399. Ward Richardson had also pioneered the localised application of ether through a spray in 1866 but although the surgeon Thomas Bryant had found some benefit from its use in opening abscesses, and removing small tumours or external piles, Charles Bell Taylor, surgeon to the Nottingham and Midland Eye Infirmary, saw it employed in the ophthalmic clinic at Guy's and described it as a 'total failure'. Bryant, and *L* II (1872) 879.
40. *L* I (1872) 123–4.
41. *L* II (1872) 241–2.
42. Ibid.
43. De Ville 1990.
44. *L* II (1872) 241–2.
45. Ibid.
46. Ibid.
47. G/S1/164, *BMJ* I (1873) 62.
48. Berkeley Hill, professor of clinical surgery at University College, produced a new edition of his manual on bandaging specifically to include reference to ether. Hill 1876. In 1879 a correspondent to the *Lancet* wrote that: 'ether is at present used in every London and many provincial hospitals', *L* II (1879) 540.
49. *L* II (1872) 879–80.
50. *BMJ* I (1876) 74–5.
51. Quoted in Duncum 1994, pp. 192–3.
52. Duncum 1994, pp. 247–9.
53. Lawrence 1987, p. 31.
54. Quoted in Duncum 1994, p. 538.
55. Ibid., pp. 538–9.
56. Ibid., pp. 539–41.
57. *L* I (1882) 364.
58. *MTG* 4 (1852) 349.
59. Duncum 1994, p. 604.
60. Ibid., p. 604.
61. *OC* 299, 316 and Duncum 1994, p. 298.
62. *BMJ* I (1877) 132–3.
63. *OC* 260–1.
64. See account of chloroform fatality at University College hospital in 1892 during which Dudley Buxton who gave the anaesthetic took control of resuscitation procedures. *L* II (1892) 1130–1.
65. Duncum 1994, pp. 378–9.
66. Bernard 1875, p. 178.
67. By 1870s it was in routine use at St Bartholomew's, Middlesex and St Thomas' hospitals. See *L* I (1863) 148, *BMJ* I (1876) 13.
68. Atropine was a narcotic which had been in pharmacopoeias since the 1830s and was used in ophthalmology to treat excessively dilated pupils. *L* II (1891) 843.

69. Swiss surgeon Gustave Julliard published his method of using morphine before ether in the *BMJ*: 'the injection is given to the patient in a quiet room, and he is encouraged to close his eyes to sleep. In about twenty minutes he is carried to the operating table where in quietness and without excitement he is etherised', *BMJ* I (1891) 920.
70. Buxton noted that: 'a terrified patient after a sleepless night is in the worst condition for an anaesthetic and an operation. In such patients, I am convinced that the use of scopolamine and morphine injections before a general anaesthetic is valuable', quoted in Duncum 1994, p. 403. In the 1900s, scopolamine and morphine combinations were used to produce 'twilight sleep' for labouring mothers, Loudon 1992, pp. 346–8.
71. See Lawrence 1987.
72. A Miles, *Surgical ward work and nursing*, 2nd edn (London: Scientific Press, 1899), quoted in Steward 1989.
73. *L* I (1876) 163.
74. Ibid.
75. *L* II (1891) 432. See also Burney 2000, p. 162.
76. Underwood 1885, p. 80. He also advised that a woman servant should chaperone the patient to ensure the absence of friends of the patient which 'contributes much to the ease of the operator and the success of the operation'.
77. Hewitt 1886, p. 97.
78. Ibid.
79. *British Journal of Dental Science* 11 (1869) 528–9.
80. Underwood 1885, pp. 79–80.
81. *LMG* 8 (1849) 455.
82. Merrington 1976, p. 57.
83. See report by London dental surgeon, C. S. Tomes, who visited Massachusetts General Hospital in 1873 and witnessed ether administrations. He noted the 'main force' by which surgeons held down a patient 'till he succumbs' to the influence of ether. So much ether was used that Tomes noted how it ran down the patient's face and neck. *BMJ* I (1873) 297.
84. Osborn 1881, p. 7.
85. H9/GY/D40/1–33.
86. Yeo pp. 70, 73 and *St Bartholomew's Hospital Reports* (1886) 51. London was not alone in experiencing such growth: by 1870 Edinburgh manufacturers of chloroform were making around 8000 doses a day – around 2.5 million doses a year – which Simpson claimed as firm evidence of how the practice of 'wrapping men, women and children in painless sleep during some of the most trying moments and hours of human existence' had become universal. Simpson 1870.
87. H9/GY/A264/1–2.
88. G/S1/164.
89. Fergusson 1852, p. 24.
90. *BMJ* I (1873) 196.
91. Ibid. (1870) 100, Squibb 1871.
92. Underwood 1885, p. 72.
93. *L* I (1896) 634.
94. Ibid. (1901) 64.

95. *Practitioner* 57 (1896) 387–93.
96. *CB* 12.11.51.
97. Underwood 1885, p. 71.
98. Sansom 1865, p. 132.
99. Quoted in Howat 2003. See also Trombley 1989, pp. 124–31.
100. Beinart 1987, Stevens 1966.
101. *OC* xl.
102. Peterson 1978, p. 207 and William Fergusson *Daybook*, 1858, held by the Royal College of Surgeons of England.
103. Zuck 2004.
104. Howat 2004.
105. *Practitioner* 57 (1896) 387–93.
106. Ibid., 379.
107. *BMJ* I (1876) 74.
108. *L* I (1883) 837–8.
109. *L* II (1900) 1908.
110. Duncum 1994, p. 539.
111. *L* I (1892) 1178–80.
112. Burney 2000.
113. Peterson 1978.
114. Digby 1994.
115. Desmond and Moore 1991, p. 371.
116. Lawrence and Treasure 2003.
117. Worboys 2000, pp. 7, 292.
118. Geison 1978.
119. Bernard 1999, p. 136.
120. *Edinburgh Medical Journal* 10 (1864–65) 190.
121. Ward Richardson 1897, p. 307.
122. Buxton 1897.
123. *L* II (1897) 1376. The Hyderabad Commission was instigated by Edward Lawrie, an old pupil of Syme's who had become surgeon to the Indian state of Hyderabad and director of the medical school. He taught his students the Scottish method of giving chloroform and used a Royal visit by the Duke and Duchess of Connaught in 1889 to attack the London method of giving chloroform by detailing experiments carried out in Hyderabad which confirmed respiration as the area of risk, rather than the heart. The event attracted such attention that the *Lancet* paid for Dr Thomas Lauder Brunton, Edinburgh graduate and physician at St Bartholomew's, to go to Hyderabad and repeat the experiments. One of the reasons Brunton was chosen was that he had already published his view that chloroform killed through direct action on the heart, rather than the respiratory death supported by the Scottish method. Within weeks, Lawrie convinced Brunton to change his view and Brunton wrote to the *Lancet* confirming Lawrie's claim that the danger of chloroform was respiratory depression, not direct action on the heart. An 1892 report on anaesthetic practice in Edinburgh written by American visitor Laurence Turnbull noted that 'prior to all operations in the Royal Infirmary, the heart, lungs and kidneys are examined... the chloroform is administered on a towel by a senior student who has been under instruction for six weeks'. Quoted in Duncum 1994, pp. 541–2.

124. *L* II (1897) 1376.
125. Worboys 2000, p. 288.
126. *L* II (1897) 1376.
127. In 1908 the Society amalgamated with the Royal Society of Medicine and became the Section of Anaesthetists.
128. *L* I (1903) 147.
129. Ibid. 81.
130. Quoted in Duncum 1994, p. 547.
131. Lawrence 1985, p. 510.

## Conclusion: The history of anaesthesia

1. A R Aitkenhead, 'Anaesthetic disasters: Handling the aftermath', *Anaesthesia*, 52 (1997) 477–82, quoted in S White, 'Death on the table', *Anaesthesia*, 58 (2003) 515–9.
2. Bernard 1875, p. 31.
3. See comments of Snow in preface to *On Ether*:

> the power we have acquired... of inducing at will and with perfect safety such a state of insensibility as we should previously have thought alarming, cannot be without its influence on the progress of our knowledge of diseases of which insensibility forms a symptom, and of the functions of the nervous system generally. *OE* 1847, pp. v–vi.

4. *L* I (1865) 613–17.
5. Ackernecht 1953, Brieger 1992, Trohler 1993.
6. Lawrence 1992, p. 2.
7. Fergusson 1852, p. 22.
8. Pernick 1985, pp. 223–8.
9. *L* II (1864) 143, 148. See also Bowman 1882.
10. Quoted in Bending 2000, p. 69.
11. F Treves, *The Elephant Man and Other Reminiscences* (London, New York: Cassell, 1923), p. 54 quoted in Bending 2000, p. 63.
12. Syme 1862, pp. 8–10.
13. *L* II (1864) 144.
14. In the following discussion, I draw particularly on the work of French 1975 and Bending 2000.
15. *Times*, January 29, 1847, page 6, column f.
16. French 1975, p. 30.
17. Ibid., p. 32.
18. For detailed analyses of this period, see French 1975 and Rupke 1987.
19. L Carroll, 'Some popular fallacies about vivisection', *Fortnightly Review*, 17 (1875) 854, quoted in Richards 1992, pp. 167–8.
20. Martineau 2003.
21. Bending 2000, pp. 2–3.
22. Ibid., pp. 6–7.
23. Ibid., pp. 28–30.
24. Ibid., p. 2.

25. Ibid., p. 8.
26. Jalland 1996, pp. 81–6.
27. Ibid., pp. 82–3.
28. Bending 2000, p. 241. See also Garland 1985.
29. Wiener 1994.
30. Bernard 1875, p. xix.
31. Lawrence 1985.
32. Kidd 1867, p. 7.
33. *Practitioner* 58 (1897) 352.
34. On the Hyderabad Commission, see Duncum 1994, pp. 432–5, Sykes 1992, Vol. III, pp. 199–265.
35. Lawrence 1992.
36. For example Shortt 1983, Warner 1980.
37. Pickstone 2005.
38. On the low status of anaesthesia during the first half of the twentieth century, see Beinart 1987, Stevens 1966.
39. Duncum 1994, p. 535.
40. On the development of American anaesthesia, see Bunker 1972, Kane, Milne and Smith 2003, Kane and Smith 2004.
41. There were occasional instances in Britain when nurses were used to give anaesthetics. The Norfolk and Norwich Hospital, for example, employed nurse anaesthetists – under the control of honorary anaesthetics – to cope with the shortage of anaesthetists at the time of the First World War. Woollam 2002.
42. Kane and Smith 2004, p. 796.
43. Bunker 1972, pp. 52, 57.
44. Kane and Smith 2004, p. 797.
45. Ibid., p. 795.
46. Kane and Smith record a series of studies which attempt to differentiate outcome of anaesthetics against each type of provider – none have provided conclusive evidence. Kane and Smith 2004, pp. 798–9.
47. Kane and Smith 2004, p. 800.
48. Kane, Milne and Smith 2003, p. 32.
49. Kane and Smith 2004, p. 796.
50. Kane, Milne and Smith 2003, p. 163.
51. Ibid.

# Bibliography

## Archival sources

### King's College Hospital Archives


*King's College Hospital*

KCH/CN/62–113 Casenotes of William Fergusson, 1840–78.
Hospital Annual Reports.

*Guy's Hospital Medical Society*

G/S1/22–30 Surgical casereports.
G/S1/164 Surgical casereports.


### Liverpool Medical Institute


Bickersteth, E R, 'On the mode of death from the inhalation of chloroform', September 1853, reprint from the Monthly Journal of Medical Science.


### London Metropolitan Archives


*Guy's Hospital*

H9/GY/B3/1–16 Registers of admissions and discharges of surgical cases.
H9/GY/A262/1 1863 Tables of patients treated at Guy's.
H9/GY/A264/1 Monthly and yearly return of anaesthetics, 1908–12.
H9/GY/A286/1–2 Lists of surgeons and physicians appointed to the hospital, 1725–1894.

*Westminster Hospital*

Annual Reports


### The Medical Society of London


Archives of the Westminster Medical Society including:
Minutes
Attendance book 1837–1842.


### The Royal College of Physicians


The Casebooks of Dr John Snow (1848–58), 3 vols.


### The Royal College of Surgeons of England


MS Fergusson, William, *Day Books*, 1847–60.
MS Brodie, Benjamin Collins, *Lecture Notes on Poisonous Gases*, Vol. I, 1821.

### Royal Free Hospital Archives

Annual Medical Reports

### Royal Society of Medicine of London

MS Papers of the Royal Medical and Chirugical Society and the Royal Society of Medicine.

### St Bartholomew's Hospital Archives and Museum

MS Medical and surgical casereports.
St Bartholomew's Hospital Reports

### St George's Hospital Archives

Hospital Minute Book, 8 July 1846–15 October 1850.
Annual Reports.

### Wellcome Library

MS7207 William Henry Ackland, Medical notebook and recipe book.
MS5461 Joseph Clover and Dudley Wilmot Buxton, Correspondence and papers relating to anaesthesia.

### Woodward Biomedical Library, University of British Columbia

MS Clover/Snow Collection

## Primary sources

Agnew, D. Hayes, *The Principles and Practice of Surgery Being a Treatise on Surgical Diseases and Injuries*, 3 vols (Philadelphia and London: J B Lippincott & Co, 1878).
Allingham, Herbert W, *The Diagnosis and Treatment of Diseases of the Rectum* (London: Churchill, 1888),
Anstie, Francis E., 'State Medicine', *Macmillan's Magazine*, XI (February, 1865).
Arnott, James, *On the Treatment of Cancer by the Regulated Application of an Anaesthetic Temperature* (London: John Churchill, 1851).
——, *On Benumbing Cold as a Preventive of Pain and Inflammation from Surgical Operations with Minute Directions for Its Use* (London: John Churchill, 1854a).
——, *The Question Considered; Is It Justifiable to Administer Chloroform in Surgical Operations, After Its Having Already Proved Suddenly Fatal in Upwards of Fifty Cases When Pain Can Be Safely Prevented Without Loss of Consciousness By Momentary Benumbing Cold?* (London: John Churchill, 1854 b).
——, *On The Expediency of Instituting an Academy of Medicine in England; Illustrated by the Prevailing Opinions and Practice Respecting the Use of Chloroform in Operations* (London: John Churchill, 1857).
Ashurst, John (ed.), *The International Encyclopaedia of Surgery*, 6 vols (London: Macmillan and Co., 1882).
Austen, Jane, *Sense and Sensibility* (New York: Harper & Row, 1961).
Bainbrigge, W. H., *Remarks on Chloroform Alleviating Human Suffering, Addressed Particularly to the Female Sex Showing That There Is No Scriptural Authority to Interdict*

*Its Administration for the Mitigation of Physical Pain Under Any Circumstances* (London: S. Highley, 1848).
Beatty, Thomas, 'Chloroform in parturition', *Dublin Quarterly Journal of Medicine* (May 1854).
Beckett, William, *A Collection of Chirugical Tracts* (London: E. Curll, 1740).
Bell, Charles, *The Letters of Charles Bell* (London: J. Murray, 1870).
Bell, Charles, *Idea of a New Anatomy of the Brain* (London: Strahan and Preston, 1811).
Bell, Joseph, *A Manual of the Operations of Surgery* (London: Robert Hardwicke, 1869).
Bernard, Claude, *Experimental Medicine*, trans. Henry Copley Greene (New Brunswick, London: Transaction Publishers, 1999).
——, *Lectures on Anaesthetics and on Asphyxia*, trans. B. Raymond Fink (Park Ridge, Illinois: Wood Library-Museum of Anaesthesiology, 1989).
Bichat, Xavier, *Physiological Researches on Life and Death* (New York: Arno Press, 1977).
Bird, James, *What to Observe in Medicine, or the Means of Improving it as a Science and an Art, with the Duties of the Medical Professional to their Patients, the Public and themselves*, An introductory address to the Harveian Medical Society (London, 1853).
Black, Patrick, *Chloroform: How Shall We Ensure Safety in Its Administration?* (London: John Churchill, 1855).
Bland Sutton, John, *The Story of a Surgeon* (London: Methuen, 1930).
Bowman, William, *The Address in Surgery, Read at Chester, August 9 1866, at the 24th Annual Meeting of the British Medical Association* (London: J. W. Kolckmann, 1882).
Braid, James, *Neurypnology, or; The rationale of Nervous Sleep Considered in Relation to Animal Magnetism* (London: John Churchill, 1843).
Brookes, W. P., *Practical Remarks on the Inhalation of the Vapour of Sulphuric Ether* (London: John Churchill, 1847).
Bryant, Thomas, *The Practice of Surgery, A Manual* (London: J. A. Churchill, 1872).
Buxton, Dudley Wilmot, *Anaesthetics: Their Uses and Administration* (London: H. K. Lewis, 1888).
——, 'Sleep and her twin sister, death', *Contemporary Review* (January 1913) 103–108.
Channing, Walter, *Cases of Inhalation of Ether in Labor* (Boston: White and Potter, 1847).
Christison, R. A., *A Treatise on Poisons* (Edinburgh: Adam Black, 1845).
Clendon, J. Chitty, *On the Use of Chloroform in Dental Surgery* (London: S. Highley, H. Bailliere, 1849).
——, *A Few Remarks on the Mode of Administering Chloroform* (London: S. Highley, H. Bailliere, 1855).
Clarke, J. F., *Autobiographical Recollections* (London: J. and A. Churchill, 1874).
Clover, J. T., 'The administration of chloroform', *BMJ* 1 (1870) 251.
Coates, W. Martin, *On Chloroform and Its Safe Administration* (London: Walton and Maberly, 1858).
Coleman, A., *Anaesthesia Considered Especially in Reference to Operations in Dental Surgery* (London: J. E. Adlard, 1862).
——, *Manual of Dental Surgery and Pathology* (London: Smith, Elder, 1881).
Collins, Robert, *A Short Sketch of the Life and Writings of the Late Joseph Clarke esq. MD, Results of His Private Practice Extending Over a Series of Forty-Four Years Including 3,878 Births* (London: Longman, Brown, Green and Longmans, 1849).

Cooper, Bransby B., *Lectures on the Principles and Practice of Surgery* (London: John Churchill, 1851).
Cooper, Samuel, *Dictionary of Practical Surgery* (London: Longman 1822, 1829, 1838, 1872).
Crombie, John M., *The Induction of Sleep and Insensibility to Pain by the Safe Self-Administration of Anaesthetics* (London: J. and A. Churchill, 1876).
Curling, T. B., *The Advantages of Ether and Chloroform in Operative Surgery, An Address Delivered to the Hunterian Society on 9 February 1849* (London: S. Highley, 1848).
Curry, James, *Popular Observations on Apparent Death from Drowning, Suffocation, etc with an Account of the Means to be Employed for Recovery* (London: Northampton, 1792).
Dalton, John, 'Experimental enquiry into the proportion of the several gases or elastic fluids, constituting the atmosphere', *Memoirs of the Literary and Philosophical Society of Manchester*, 1 (1805) 244–58.
Davis, Henry, *Guide to the Administration of Anaesthetics* (London: H. K. Lewis, 1887).
Davy, Humphry, *Researches, Chemical and Philosophical; Chiefly Concerning Nitrous Oxide, or Dephlogisticated Nitrous Air and its Respiration* (London: Butterworths, 1972).
Doyle, Arthur Conan, *The Complete Adventures of Sherlock Holmes*, ed. Julian Symons (America: Martin Secker and Warburg Ltd, 1981).
Druitt, Robert, *The Surgeon's Vade Mecum, A Handbook of the Principles and Practice of Surgery* (London: Renshaw, 1854).
——, *The Principles and Practice of Modern Surgery*, ed. F. W. Sargent (Philadelphia: Blanchard and Lea, 1852).
Duroy, M., *The Anaesthesimeter: A New Apparatus for the Administration of Chloroform* (Brighton: Richard Sickelmorem, 1856).
Elliotson, John, *The Principles and Practice of Medicine* (London: J. Butler, 1839).
——, *Harveian Oration... with an English version and notes* (London: H. Bailliere, 1846).
Fergusson, William, *A System of Practical Surgery* (London: J. Churchill, 1842, 1853, 1870).
Flourens, Marie Jean-Pierre, *Experimental Researches on the Properties and Functions of the Nervous System in Vertebrates* (Paris: Crevot, 1824).
Freeman, R. T., *A Few Words Upon Anaesthetics*, (London: J. and A. Churchill, 1883).
Goodwyn, Edmund, *The Connexion of Life with Respiration* (London: Lauren J. Johnson, 1788).
Gream, G. T., *Remarks on the Employment of Anaesthetic Agents in Midwifery* (London: John Churchill, 1848).
Guthrie, George J., *Commentaries on the Surgery of the War in Portugal, Spain, France, and the Netherlands* (London: H. Renshaw, 1853, 1855).
Halford, Sir, Henry, *On the Education and Conduct of a Physician* (London: J. Murray, 1834).
Halton, John P., *The Results of the Great Operations of Surgery, During a Practice of Twenty-Two Years, Showing a High Rate of Success at the Liverpool Infirmary* (Liverpool, 1843).
Herschel, John F. W., *Preliminary Discourse on the Study of Natural Philosophy* (London: Longman, Brown, Green and Longmans, 1851).
Hewitt, Frederic, *The Selection and Administration of Anaesthetics with Remarks Concerning Certain Forms of Respiratory Failure* (1886).

Hickman, Henry Hill, *A Letter on Suspended Animation, Containing Experiments Shewing that It May Be Safely Employed During Operations on Animals, With the View of Ascertaining Its Probable Utility in Surgical Operations on the Human Subject, Addressed to T. A. Knight, Esq.* (Ironbridge: W. Smith, 1824).

Hill, Berkeley, *The Essentials of Bandaging: With Directions for Managing Fractures and Dislocations; For Administering Ether and Chloroform and for Using Other Surgical Apparatus* (London: Smith, Elder and Co., 1876).

Home, Francis, *Medical Facts and Experiments* (London: A. Miller, 1759).

Holmes, Timothy, *Sir Benjamin Collins Brodie* (London: T. Fisher Unwin, 1898).

*Hooper's Dictionary* (London: Henry Renshaw, 1798, 1820, 1839, 1848).

Hudson, J. C., *The Parent's Handbook* (London, 1842).

Hunter, John, 'Proposals for the recovery of people apparently drowned', *Philosophical Transactions of the Royal Society of London*, XXIV (1776) 412–25.

Imlach, Francis Brodie, *On the Employment of Chloroform in Dental Surgery, Its Mode of Exhibition etc* (Edinburgh: Sutherland & Knox, 1848).

James, J. H., *Chloroform Versus Pain* (London: John Churchill, 1870).

Jeaffreson, J C, *A Book About Doctors* (London: Hurst and Blackett, 1861).

Johnson, A., *A Short Account of a Society at Amsterdam Instituted in the Year 1767 for the Recovery of Drowned Persons* (London, 1773).

Keith, William, *On the Safety and Suitableness of Chloroform as an Anaesthetic Agent in Some of the More Complex and Serious Operations in Surgery* (Edinburgh: Sutherland and Knox, 1849).

Kidd, Charles, *On Aether and Chloroform as Anaesthetics* (London: Renshaw, 1858).

——, *On Chloroform* (London: Renshaw, 1867).

*Kilvert's Diary, 1870–79* (London: Century Publishing, 1986).

Kirk, Robert, *A New Theory of Chloroform Syncope Showing How the Anaesthetic Ought to be Administered* (Glasgow: John Thomlinson, 1890).

*Knight's Cyclopaedia of London* (London: Charles Knight, 1851).

Laing Gordon H., *Sir James Young Simpson and Chloroform* (New York: Longmans, Green & Co, 1897).

*Laws (The) of the Westminster Medical Society Instituted AD1809, For the Advancement of Medical Science* (London, 1835).

Liston, Robert, *Practical Surgery* (London: Churchill, 1838).

Lizars, John, *A system of practical surgery* (London: S. Highley, 1847).

Lloyd, Rickard W., *The Field View of the Anaesthetist* (London: John Bale and Sons and Danielsson, 1902).

Macartney, James, *A Treatise on Inflammation* (London: Longman and Co., 1838).

Macmichael, William, *Lives of British Physicians* (London: J. Murray, 1830).

Martineau, Harriet, *Life in the Sick Room*, ed. Maria Frawley (Canada: Broadview, 2003).

Merriman, S. W. J., *Arguments Against the Indiscriminate Use of Chloroform in Midwifery* (London: John Churchill, 1848).

Mill, John Stuart, *On Liberty and Other Essays* (Oxford: Oxford University Press, 1991).

Miller, James, *Surgical Experience of Chloroform* (Edinburgh: Sutherland and Knox, 1848).

——, *The Practice of Surgery* (London: Longman and Co., 1852).

——, *The Principles of Surgery* (London: Longman and Co., 1853).

Morgan, J, *The Dangers of Chloroform and the Safety and Efficiency of Ether* (London: Balliere, Tindall and Cox, 1872).

Murphy, Edward, *Chloroform in the Practice of Midwifery* (London: Taylor and Walton, 1848).
——, *Further Observations on Chloroform in the Practice of Midwifery* (London: Taylor, Walton and Maberly, 1848).
——, *Chloroform: Its Properties and Safety in Childbirth* (London: Walton and Maberly, 1855).
Neill, Hugh, *On the Cure of Cataract, With a Practical Summary of the Best Modes of Operating, Continental and British* (Liverpool: Deighton and Laughton, 1848).
Samuel Osborn, *Annotations on Anaesthetics* (London: J. E. Adlard, 1881).
Paget, Sir James, *Memoirs and Letters of Sir James Paget*, Stephen Paget, ed. (London: Longmans, Green & Co., 1901).
Peacock, George, *The Life of Thomas Young, MD, FRS* (London: J. Murray, 1855).
Percival, Thomas, *Medical Ethics* (Birmingham Alabama: Classics of Medicine Library, 1985).
Poore, Vivian, 'Clinical Remarks on Chloroform and its Administration', *Lancet* II (1872) 521–1.
Prichard, Augustin, *A Few Medical and Surgical Reminiscences* (Bristol: J. W. Arrowsmith, 1896).
Quinton, J. Richard, *Painless Tooth Extraction by Congelation* (London: Ward & Co, 1856).
Richardson, Benjamin Ward, 'John Snow MD, A representative of medical science and art of the Victorian era', *Asclepiad* 4 (1887) 274–300.
——, *Biological Experimentation: Its Functions and Limits* (London: G. Bell & Sons, 1896).
——, *Vita Medica* (London: Longmans, Green and Co., 1897).
Robinson, James, *On the Inhalation of the Vapour of Ether,* With a preface by Richard H. Ellis (Eastbourne: Bailliere Tindall, 1983).
Peirera, Jonathon, *Elements of Materia Medica* (London: Longman, 1843).
Sansom, A. E., *Chloroform: Its Action and Administration, A Handbook* (London: John Churchill, 1865).
Simpson, J. Y., *Ovariotomy; Is It – Or Is It Not – An Operation Justifiable Upon the Common Principles of Surgery? Are – Or Are Not – Capital Operations in Surgery Justifiable to the Extent Generally Practised?* (Edinburgh: Sutherland & Knox, 1846).
——, *Account of a New Anaesthetic Agent as a Substitute for Sulphuric Ether in Surgery and Midwifery* (Edinburgh: Sutherland & Knox, 1847).
——, *Remarks on the Superinduction of Anaesthesia* (Boston: W. B. Little & Co., 1848).
——, *Answer to the Religious Objections Advanced Against the Employment of Anaesthetic Agents in Midwifery and Surgery* (Edinburgh: Sutherland & Knox, 1848b).
——, *Anaesthesia or the Employment of Chloroform and Ether in Surgery, Midwifery etc* (Philadelphia: Lindsay and Blakiston, 1849).
——, *Anaesthesia, Hospitalism and Other Papers* (Edinburgh: Sutherland & Knox, 1871).
Smith, John, *The Present Position of Dental Surgery and Its Advancement as a Branch of Medical Education, A Letter Addressed to Dr Robert Omond, President of the Royal College of Surgeons* (Edinburgh: Sutherland & Knox, 1858).
Smith, Mayo G., *A Treatise on the Inhalation of Ether for the Prevention of Pain,* (Boston: J. P. Jewett, 1848).
Smith, Mayo G., *A Brief Treatise on the Use of Chloroform in Dental and Surgical Operations* (Boston, 1849).

Snow, John, 'Arsenic as a preservative of dead bodies.' *L* I (1838–39) 264.
——, 'Action of the recti muscles.' *LMG* 23(1838–39) 559–60.
——, 'On the bands in the recti muscles.' *LMG* 23 (1838–39) 719–20.
——, 'Mechanism of respiration.' *L* I (1838–39) 653–5.
——, 'On the effects of carbonic acid.' *LMG* 24 (1839–40) 60–2.
——, 'On distortions of the chest and spine in children from enlargement of the abdomen.' *LMG* 28 (1840–41) 112–16.
——, 'On asphyxia, and on the resuscitation of still-born children.' *LMG* 29 (1841–42) 222–7.
——, 'On paracentesis of the thorax.' *LMG* 29 (1841–42) 705–7.
——, 'On the circulation in the capillary blood-vessels, and on some of its connections with pathology and therapeutics.' *LMG* 31 (1842–43) 810–16.
——, 'Case of acute poisoning by carbonate of lead.' *LMG* 35 (1844–45) 248–50.
——, 'On the use of the term "Allopathy"'. *L* I (1846) 229.
——, 'On the pathological effects of atmospheres vitiated by carbonic acid gas and by a diminution of the due process of oxygen.' *Edinburgh Medical and Surgical Journal* 65 (1846) 49–56.
——, 'Table for calculating the strength of ether vapour.' *LMG* 39 (1847) 219–20.
——, 'Observations on the vapour of ether, and its applications to prevent pain in surgical operations.' *L* I (1847) 227–8.
——, 'On the inhalation of the vapour of ether.' *LMG* 39 (1847) 498–502 and 539–42.
——, 'The inhalation of vapour of ether in surgical operations.' *L* I (1847) 551–4.
——, *On the Inhalation of the Vapour of Ether in Surgical Operations: Containing a Description of the Various Stages of Etherisation and a Statement of the Results of Nearly Eighty Operations in Which Ether Has Been Employed* (London: Churchill, 1847).
——, 'On the inhalation of chloroform and ether.' *L* I (1848) 177–80 and *LMG* 6 (1848) 74–6.
——, 'The fatal chloroform case at Newcastle.' *L* I (1848) 239.
——, 'Remarks on the fatal case of inhalation of chloroform.' *LMG* 6 (1848) 277–8.
——, 'On narcotism by the inhalation of vapours.' *LMG* (1848–51).
——, *On Narcotism by the Inhalation of Vapours*, Richard H. Ellis ed. (London: Royal Society of Medicine, 1991).
——, 'On the use of chloroform in surgical operations and midwifery.' *London Journal of Medicine* I (1849) 50–55.
——, 'On the discussion respecting chloroform in the Académie de Medécine de Paris.' *London Journal of Medicine* I (1849) 324–6.
——, 'On the fatal cases of inhalation of chloroform.' *Edinburgh Medical and Surgical Journal* 72 (1849) 75–8.
——, 'The alleged employment of chloroform by thieves.' *LMG* 10 (1850) 327.
——, 'Further remarks on the employment of chloroform by thieves.' *LMG* 11 (1850) 834–51.
——, 'On the inhalation of various medicinal substances.' *London Journal of Medicine* 3 (1851) 122–9.
——, 'Chloroform and its use by thieves.' *LMG* 12 (1851) 571.
——, *Letter to the Right Honourable Lord Campbell, Lord Chief Justice of the Court of Queen's Bench, on the Clause Respecting Chloroform in the Proposed Prevention of Offences Bill* (London: John Churchill, 1851).

——, 'On the administration of chloroform in the public hospitals.' *MTG* 4 (1852) 349–50.
——, 'On the cause and prevention of death from chloroform.' *London Journal of Medicine* 4 (1852) 320–9, 415–23, 564–72.
——, 'Deaths from chloroform in Scotland.' *MTG* 4 (1852) 598–9.
——, 'Death from chloroform in a case of fatty degeneration of the heart.' *MTG* 5 (1852) 361–2.
——, 'On the administration of chloroform during parturition.' *AMJ* I (1853) 500–2.
——, 'The late deaths from chloroform.' *MTG* 7 (1853) 485.
——, 'On the modus operandi of narcotico-irritants. *L* II (1853) 586 and *MTG* 7 (1853) 641–2.
——, 'On the recent death at St George's Hospital, while the patient was beginning to inhale chloroform.' *MTG* 8 (1854) 606.
——, 'Chloroform in London and Edinburgh.' *L* I (1855) 108.
——, 'Employment of chloroform in surgical operations.' *L* II (1855) 361–3, 383–5.
——, 'The breathing and the pulse under the influence of chloroform.' *AMJ* 3 (1855) 313–18.
——, 'Further remarks on the cause and prevention of death from chloroform.' *L* I 1856) 148–50.
——, 'Administration of chloroform'. *MTG* 13 (1856) 21.
——, 'On the vapour of amylene.' *MTG* 14 (1857) 60–2, 80–2.
——, 'On the recent accident from chloroform.' *MTG* 14 (1857) 282–3.
——, 'Further remarks on amylene.' *MTG* 14 (1857) 332–4, 357–9, 379–82.
——, 'On some alleged ill-effects of amylene.' *BMJ* I (1857) 381.
——, 'Dr Marshall Hall's method of artificial respiration. *MTG* 14 (1857) 421.
——, 'On chloride of amyl.' *MTG* 14 (1857) 457.
——, 'Mr A Prichard on amylene.' *BMJ* 2 (1857) 654.
——, 'Case of death from amylene.' *MTG* 15 (1857) 133–4.
——, 'The recent death from chloroform at Bristol.' *BMJ* (1858) 223–4.
——, 'Death from chloroform.' *BMJ* 1 (1858) 279.
——, *On Chloroform and Other Anaesthetics: Their Action and Administration* (London: J. Churchill 1858).
Sprigge, Samuel Squire, *The Life and Times of Thomas Wakley* (London: Longmans, Green & Co., 1897).
——, *Medicine and the Public* (London: William Heinemann, 1905).
Squibb, Edward R., *Anaesthetics* (New York, 1871).
Squire, William, 'On the introduction of ether inhalation as an anaesthetic in London', *Lancet* II (1888) 1220–1.
Stowe, William, *A Manual of toxicology; In Which the Symptoms, Treatment, and Tests of the Various Poisons, Mineral, Vegetable, and Animal, are Concisely Stated. To Which are Added, Directions for the Recovery of Persons in a State of Suspended Animation.* (London: John Anderson, 1823).
Tracy, Samuel J., *A Description of an Apparatus for the Inhalation of Ether Vapour; with Some Remarks on Its Use* (London: D. Ferguson, 1847).
*Transactions of the Society of Anaesthetists* (London: Medical Publishing Company Limited, 1898–1908).
Treves, Frederick, *A Manual of Surgery*, 3 vols (London: Cassell & Co., 1886).
Tweedie, Mrs Alec, *George Harley FRS, The Life of a London Physician* (London: Scientific Press, 1899).

Underwood, Arthur S., *Notes on Anaesthetics with an Appendix Containing Illustrative Cases and Engravings of an Anaesthetic Apparatus* (London: Claudius Ash, 1885).
Waldie, D., *The True Story of the Introduction of Chloroform into Anaesthetics* (Edinburgh: Linlithgow, 1870).
Warren, J. C., *Effects of Chloroform and of Strong Chloric Ether* (Boston: W. D. Ticknov, 1849).
Warren, J. Mason, *Surgical Observations with Cases and Operations* (Boston: Ticknor and Fields, 1867).
Whiter, Walter, *A dissertation on the Disorder of Death; Or that State of the Frame Under the Signs of Death Called Suspended Animation;... in Which it is Recommended, that the Same Remedies of the Resuscitative Process Should be Applied...&c* (Norwich: S. Hayes, 1819).
Wilson, *Memoir of George Wilson MD FRSE, By His Sister* (London and Cambridge: Macmillan, 1862).

## Secondary sources

Abel Smith, B., *The Hospitals* (London: Heiremann, 1964).
Ackerknecht, E. H., *Rudolf Virchow: Doctor, Statesman, Anthropologist* (Madison: University of Winsconsin Press, 1953).
——, *Medicine at the Paris Hospital, 1794–1848* (Baltimore: John Hopkins University Press, 1967).
——, *A Short History of Medicine*, revised edition (Baltimore and London: John Hopkins University Press, 1982).
Albury, W. R., 'Ideas of Life and Death', in Bynum and Porter, vol. 1 (1993) 249–80.
Aries, Philippe, *The Hour of Our Death* (Oxford: Oxford University Press, 1991).
Armstrong Davison, M. H., *The Evolution of Anaesthesia* (Altrincham: John Sherratt and Son, 1965).
Atkinson, Paul, *Medical Talk and Medical Work* (Cambridge: SAGE publications, 1995).
Atkinson, Richard S., *James Simpson and Chloroform* (London: Priory Press, 1973).
Atkinson, Richard S. and Thomas B. Boulton, *The History of Anaesthesia* (London: Royal Society of Medicine Services Limited, 1989).
Baker, Robert, Dorothy Porter and Roy Porter (eds), *The Codification of Medical Morality, Historical and Philosophical Studies of the Formalization of Western Medical Morality in the Eighteenth and Nineteenth Centuries*, vol. 1 (Dordrecht, Boston, London: Kluwer Academic Publishers, 1993).
Barnes, Barry and Steve, Shapin (eds), *Natural Order: Historical Studies of Scientific Culture* (Beverley Hills, London: Sage Publications, 1979).
Baker, Robert (ed.), *The Codification of Medical Morality, Historical and Philosophical Studies of the Formalization of Western Medical Morality in the Eighteenth and Nineteenth Centuries*, vol. 2 (Dordrecht, Boston, London: Kluwer Academic Publishers, 1995).
Barker, Juliet, *The Brontes* (London: Weidenfeld and Nicolson, 1994).
Barr, A. Marshall, Thomas B. Boulton and David J. Wilkinson (eds), *Essays on the History of Anaesthesia* (London: Royal Society of Medicine Press, 1996).
Baxby, Derrick, 'Jenner, Edward (1749–1823)', *Oxford Dictionary of National Biography* (Oxford: Oxford University Press, 2004).

Bazanger, Isabelle, *Inventing Pain Medicine: From the Laboratory to the Clinic* (New Brunswick, Jersey, London: Rutgers University Press, 1998).

Bedford, Joanne, *Who Should Deliver Babies? Models of Nature and the Midwifery Debate, 1800–1896*, unpublished PhD thesis (Birkbeck College, London, 1995).

Beecher, Henry K. and Donald P. Todd, *A Study of the Deaths Associated with Anaesthesia and Surgery* (Springfield, Illinois: Charles C. Thomas, 1954).

Beier, Lucinda McCray, 'Expertise and control: Childbearing in three twentieth-century working-class lancashire communities', *Bulletin of the History of Medicine*, 78 (2004) 379–409.

Beinart, Jennifer, *A History of the Nuffield Department of Anaesthetics, Oxford 1937–1987* (Oxford: Oxford University Press, 1987).

Bellot, H. H., *University College, London 1826–1926* (London: University of London Press, 1929).

Bending, Lucy, *The Representation of Bodily Pain in Late Nineteenth-Century English Culture* (Oxford: Oxford University Press, 2000).

Benjamin, Marina (ed.), *Science and Sensibility: Gender and Scientific Enquiry* (Oxford: Basil Blackwell, 1991).

Bennett, J. A., 'The eclipsed dawn of anaesthesia in Bristol', *Proceedings of the History of Anaesthesia Society* 25 (1999) 85–9.

Bergman, Norman A., *The Genesis of Surgical Anaesthesia* (Park Ridge III: Wood Library, Museum of Anesthesiology, 1998).

Berlant, J. L., *Profession and Monopoly: A Study of Medicine in the United States and Great Britain* (Los Angeles, Berkeley: University of California Press, 1975).

Berman, Morris, 'The early years of the Royal Institution 1799–1810: A re-evaluation', *Science Studies* 2 (1972) 205–40.

——, *Social Change and Scientific Organisation: The Royal Institution, 1799–1844* (London: Heinemann Education, 1984).

Berridge, Virginia and Griffith Edwards, *Opium and the People* (Yale University Press: New Haven and London, 1987).

Best, Geoffrey, *Mid-Victorian Britain, 1851–75*, (London: Fontana Press, 1985).

Bishop, P. J., *A Short History of the Humane Society* (London: The Royal Humane Society, 1974).

Bivins, Roberta E., *Acupuncture, Expertise and Cross-Cultural Medicine* (Basingstoke: Palgrave, 2000).

Boulton, T. B., 'The Department of Anaesthesia', *St Bartholomew's Hospital Journal* (July 1964) 303–6.

Boyd, Hilton, *The Age of Atonement: The Influence of Evangelicism on Social and Economic Thought, 1795–1865* (Oxford: Oxford University Press, 1988).

Brieger, Gert H., 'From conservative to radical surgery in late nineteenth-century America', in Lawrence 1992.

Briggs, Asa, *The Age of Improvement, 1783–1867* (New York: Longman, 1979).

——, *Victorian People* (Harmondsworth, Eng.: Penguin Books, 1990).

Browne, Janet, *Charles Darwin, Voyaging* (Princeton: Princeton University Press, 1995).

——, *Charles Darwin, The Power of Place* (London: Pimlico, 2003).

Browner, Stephanie, 'Ideologies of the Anaesthetic: Professionalism, Egalitarianism and the Ether Controversy', *American Quarterly*, 51 (1999) 108–43.

Bud, Robert and Gerrylynn K. Roberts, *Science Versus Practice: Chemistry in Victorian Britain* (Manchester: Manchester University Press, 1984).

Bunker, John P., *The Anesthesiologist and the Surgeon* (Little, Brown and Company: Boston, 1972).
Burney Ian A., *Bodies of Evidence: Medicine and the Politics of the English Inquest 1830–1926* (Baltimore and London: John Hopkins University Press, 2000).
——, 'Anaesthetic death and the evaluation of risk in 19th century english surgery', in Thomas Schlich and Ulrich Troêhler eds, *The Risks of Medical Innovation: Risk Perception and Assessment in Historical Context* (London and New York: Routledge, 2006).
——, *Poison, Detection and the Victorian Imagination* (Manchester: Manchester University Press, 2006).
Byatt, A. S., *Unruly Times: Wordsworth and Coleridge in Their Time* (London: Vintage, 1997).
Bynum, W. F. and V. Nutton (eds), *Essays in the History of Therapeutics* (Amsterdam-Atlanta, G.A. 1991).
Bynum, W. F., *Science and the Practice of Medicine in the Nineteenth Century* (Cambridge: Cambridge University Press, 1993).
Bynum, W. F., and Roy Porter (eds), *William Hunter and the eighteenth-century medical world* (Cambridge: Cambridge University Press, 1985).
—— (eds), *Medical Fringe and Medical Orthodoxy 1750–1850* (London: Croom Helm, 1987).
—— (eds), *Brunonianism in Britain and Europe* (Medical History Supplement No. 8, 1988).
—— (eds), *Medicine and the Five Senses* (Cambridge: Cambridge University Press, 1993a).
—— (eds), *Companion Encyclopaedia of the History of Medicine*, 2 vols (London and New York: Routledge, 1993b).
Bynum, W. F., Stephen Lock and Roy Porter (eds), *Medical Journals & Medical Knowledge: Historical Essays* (London and New York: Routledge, 1992).
Cameron, H. C., *Mr Guy's Hospital, 1726–1948* (London, New York: Longmans, Green, 1954).
Cannon, Susan Faye, *Science in Culture: The Early Victorian Period* (New York: Science History Publications, 1978).
Carr-Saunders, A. M. and P. A. Wilson, *The Professions* (Oxford: Clarendon Press 1933).
Cartwright, F. F., *English Pioneers of Anaesthesia* (Bristol: Johnwright, London: Simpkin Marshall, 1952).
——, *The Development of Modern Surgery* (London and New York: Arthus Baker 1967).
Caton, Donald, 'The secularisation of pain', *Anaesthesiology*, 62 (1985) 493–501.
——, ' "In the present state of our knowledge": Early use of opoids in obstetrics', *Anesthesiology* 82 (1995) 779–84.
——, *What a Blessing She Had Chloroform* (New Haven: Yale University Press, 1999).
Chaloner, E. J., 'Amputations at the London Hospital 1852–1857', *Journal of the Royal Society of Medicine* 94 (2001) 409–17.
Cock, F. W., 'The first major operation under ether in England', *American Journal of Surgery* 29 (1915) 98–1060.
Cohen, E. Muriel, 'Sir John Tomes (1815–95), *Dictionary of National Biography* (Oxford: Oxford University Press, 2004).

Conacher, I. D., 'Amylene, a blemish on Snow?', *Anaesthesia* 51 (1996) 155–7.

Connor, Henry, 'The use of chloroform by British Army Surgeons during the Crimean War', *Medical History* 42 (1998) 161–93.

Cooke, A. M., 'Queen Victoria's Medical Household', *Medical History* 26 (1982) 307–20.

Cooter, Roger, *The Cultural Meaning of Popular Science: Phrenology and the Organisation of Consent in Nineteenth Century Britain* (Cambridge: Cambridge University Press, 1984).

——, 'Dichotomy and denial: Mesmerism, medicine and Harriet Martineau'. In Marina Benjamin, ed. *Science and Sensibility: Gender and Scientific Enquiry* (Oxford: Basil Blackwell, 1991) 144–74.

——, *Surgery and Society in Peace and War* (Basingstoke: Macmillan, 1993).

Cope, Zachary, *The History of the Royal College of Surgeons of England* (London: Anthony Bland, 1959).

Cox, Alfred, *Among the Doctors* (London: C. Johnson, 1950).

Crosse, V. Mary, *A Surgeon in the Early Nineteenth Century: The Life and Times of John Green Crosse* (Edinburgh and London: E & S Livingstone, 1968).

Cunningham, Andrew and Nicholas Jardine (eds), *Romanticism and the Sciences* (Cambridge: Cambridge University Press, 1990).

Cunningham, Andrew and Perry, Williams (eds), *The Laboratory Revolution in Medicine* (Cambridge: Cambridge University Press, 1992).

Dally, Ann, 'Status lymphaticus: Sudden death in children from "visitation of god" to cot death', *Medical History*, 41 (1997) 70–85.

Peter, Davis, *The Social Context of Dentistry* (London: Croom Helm, 1980).

Davidson, Luke, 'Identities ascertained: British ophthalmology in the first half of the nineteenth century', *Journal of the Society for the Social History of Medicine* 9 (1996) 313–34.

De Ville, Kenneth Allen, *Medical Malpractice in Nineteenth Century America* (London and New York: New York University Press, 1990).

Defalque, R. F. and A. J. Wright, 'The short tragic life of Robert M. Glover', *Anaesthesia* 59 (2004) 394–400.

Desmond, Adrian, *The Politics of Evolution: Morphology, Medicine and Reform in Radical London* (Chicago: Chicago University Press, 1992).

Desmond, Adrian and J. R. Moore, *Darwin* (London: Michael Joseph, 1994).

Dewhurst, Jack, *Royal Confinements* (London: Weidenfield and Nicolson, 1980).

Digby, Anne, *Making a Medical Living: Doctors and Patients in the English Market for Medicine, 1720–1911* (Cambridge: Cambridge University Press, 1994).

Donnison, Jean, *Midwives and Medical Men: A History of Inter-Professional Rivalries and Women's Rights* (New York: Heinemann Educational Ltd, 1977).

Duncum, Barbara M., *The Development of Inhalation Anaesthesia* (London: Royal Society of Medicine Press, 1994).

Dupree, Marguerite, 'Other than healing: Medical practitioners and the business of life assurance during the nineteenth and early twentieth centuries', *Journal of the Society for the Social History of Medicine* 10 (1997) 79–104.

Durey, Michael, 'Medical elites, the general practitioner and patient power in Britain during the cholera epidemic of 1831/2', in Inkster and Morrell 1993.

Elliott, Philip, *The Sociology of the Professions* (London: Macmillan, 1972).

Ellis, Richard H., 'The introduction of ether anaesthesia to Great Britain No.1', *Anaesthesia* 31 (1976) 766–77.

——, 'The introduction of ether anaesthesia to Great Britain No.2', *Anaesthesia* 32 (1977) 197–208.
——, 'Robert Liston's letter to Francis Boott: Its reappearance after 135 years', *Anesthesiology* 62 (1985) 331–5.
Ellmann, Richard, *Oscar Wilde* (London: Hamish Hamilton, 1987).
English, Mary P., *Victorian Values: The Life and Times of Dr Edwin Lankester MD FRS* (Bristol: Biopress Ltd, 1990).
Eyler, J. M., *Victorian Social Medicine: The Ideas and Methods of William Farr* (Baltimore: John Hopkins University Press, 1979).
Fara, Patricia, *Sympathetic Attractions: Magnetic Practices, Beliefs and Symbolism in Eighteenth-Century England* (Princeton: Princeton University Press, 1996).
Farr, A. D., 'Religious opposition to obstetric anaesthesia: A myth?', *Annals of Science* 40 (1983) 159–77.
Fissell, M. E., *Patients, Power and the Poor in Eighteenth Century Bristol* (Cambridge: Cambridge University Press, 1991).
——, 'The disappearance of the patient's narrative and the invention of hospital medicine', in French and Wear 1992.
Foucault, Michel, *The Birth of the Clinic: An Archaeology of Medical Perception*, trans. by A. M. Sheridan Smith (New York: Random House, 1994).
Ford, John, 'A medical student at St Thomas' Hospital, 1801–2, The weekes family letters', *Medical History*, Supplement No. 7 (1987).
Forreger, Richard, 'John Snow's early research on carbon dioxide', *Anaesthesiology*, 21 (1960) 20–5.
Fox, Nicholas J., 'Scientific theory choice and social structure: The case of Joseph Lister's antisepsis, humoral theory and asepsis', *History of Science* 26 (1988) 365–97.
French, R. D., *Antivivisection and Medical Science in Victorian Society* (Princeton, NJ: Princeton University Press, 1975).
French Roger and Andrew Wear (eds), *British Medicine in an Age of Reform* (London and New York: Routledge, 1992).
Friedlander, Walter J., 'The Bigelow-Simpson controversy: Still another early argument over the discovery of anaesthesia', *Bulletin of the History of Medicine*, 66 (1992) 613–25.
Freidson, Eliot, *Profession of Medicine: A Study of the Sociology of Applied Knowledge* (New York: Harper and Row 1970).
——, 'The Reorganization of the Medical Profession', *Medical Care Review* 42 (1985) 11–35.
Galland, R. B., 'Nineteenth century amputations at the Royal Berkshire Hospital, Reading', *Annals of the Royal College of Surgeons of England* 85 (2003) 393–5.
Garland, David, *Punishment and Welfare: A History of Penal Strategies* (Aldershot: Gower Publishing, 1985).
Gauld, Alan, *A History of Hypnotism* (Cambridge: University Press, 1992).
Geison, Gerald L., *Michael Foster and the Cambridge School of Physiology* (Princeton: Princeton University Press, 1978).
Gilman, Sander L., *Making the Body Beautiful: A Cultural History of Aesthetic Surgery* (Princeton NJ: Princeton University Press, 1999).
Glendinning, Victoria, *Trollope* (London: Pimlico, 1993).
Golinski, J., *Science as Public Culture: Chemistry and Enlightenment in Britain, 1760–1820* (Cambridge: Cambridge University Press, 1992).

Graham, Harvey, *Eternal Eve* (London: Heinemann, 1950).

Granshaw, Lindsay, '"Fame and fortune by bricks and mortar": The medical profession and specialist hospitals in Britain, 1800–1948', in Granshaw and Porter, 1989.

——, 'Knowledge of bodies or bodies of knowledge? Surgeons, anatomists and rectal surgery, 1830–1985', in Lawrence 1992.

Granshaw, Lindsay and Roy Porter (eds), *The Hospital in History* (London: Routledge, 1989).

Gray, Ernest A., *By Candlelight: The Life of Dr Arthur Hill Hassall, 1817–1894*, (London: R. Hale, 1983).

Green, David G., *Working-Class Patients and the Medical Establishment* (Aldershot: Gower Publishing, 1985).

Greene, Nicholas M., 'Anesthesia and the development of surgery (1846–1896)', *Anesthesia and Analgesia* 58 (1979) 5–12.

Haakonssen, Lisbeth, *Medicine and Morals in the Enlightenment* (Amsterdam-Atlanta: G. A., 1997).

Hamilton, David, 'The nineteenth-century surgical revolution – antisepsis or better nutrition?', *Bulletin of the History of Medicine* 56 (1982) 30–40.

Hammonds, W. D., 'Crawford W. Long: Pioneer physician in anaesthesia', *Journal of Clinical Anesthesia* 5 (1993) 163–7.

Hannaway, Caroline and Ann La Berge (eds), *Constructing Paris Medicine* (Amsterdam-New York: Rodopi, 2003).

Hargreaves, Anne, 'Dentistry in the British Isles', in Hillam 2003.

Harte, Negley, *The University of London 1836–1986* (London: Atlantic Highlands, 1986).

Hayter, A., *Opium and the Romantic Imagination* (London: Faber & Faber, 1971).

Heaman, E. A., *St Mary's: The History of a London Teaching Hospital* (Montreal: McGill-Queen's University Press, 2003).

Hemlow, J. (ed.), *Journals and Letters of Fanny Burney*, 5 vols (Oxford: Clarendon Press, 1975).

Hillam, Christine, 'James Robinson (1813–1862), professional irritant and Britain's first anaesthetist', *2nd Lilian Lindsay Memorial Lecture, 18 May 1996, British Dental Association Conference, Edinburgh.*

——, (ed.), *Dental Practice in Europe at the End of the 18th Century* (Amsterdam-New York: Rodopi, 2003).

Holloway, S. W. F., 'Medical Education in England, 1830–1858: A sociological analysis', *History* XLIX (1964) 299–324.

——, 'The Apothecaries' Act, 1815: A reinterpretation', *Medical History* 10 (1966) 107–29 and 221–36.

——, 'The regulation of drugs in Britain before 1868', in Porter and Teich eds (1995).

Howarth, O. J. R., *The British Association for the Advancement of Science: A Retrospect* (London: The Association, 1922).

Howat, D. D. C., 'Frederic Hewitt's account of Edward VII's Appendicectomy', *Proceedings of the History of Anaesthesia Society* 33 (2003) 14–17.

——, 'Hewitt, Sir Frederic William (1857–1916)', *Dictionary of National Biography* (Oxford: Oxford University Press, 2004).

Howell, Joel D., *Technology in the Hospital* (Baltimore and London: John Hopkins University Press, 1995).

Howie, W. B., 'Complaints and complaint procedures in the eighteenth and early nineteenth century provincial hospitals in England', *Medical History* 25 (1981) 345–62.

Hunter, Kathryn Montgomery, *Doctors' Stories* (Princeton: Princeton University Press, 1991).

Hunt, Thomas (ed.), *The Medical Society of London, 1773–1973* (London: Heinemann Medical for the Medical Society of London, 1973).

Inkster, Ian, *Science and Technology in History: An Approach to Industrial Development* (Basingstoke: Macmillan, 1991).

Inkster, Ian and Jack Morrell (eds), *Metropolis and Province: Science in British Culture, 1780–1850* (London: Hutchinson, 1983).

Jackson, William A., *The Invention of the Stomach Pump and Its Development in the Nineteenth Century*, unpublished MSc thesis, University of Manchester, 1996.

Jacob M. C. and M. J. Sauter, 'Why did Humphry Davy and Associates not pursue the pain-alleviating effects of nitrous oxide?', *Journal of the History of Medicine and Allied Sciences* 57 (2002) 161–76.

Jacyna, L. S., 'Immanence or transcendence: Theories of life and organisation in Britain, 1790–1835', *Isis* 74 (1982) 310–329.

—— (ed.), *A Tale of Three Cities: The Correspondence of W. Sharpey and A. Thomson* (Medical History Supplement 9, 1989).

——, *Lost Words: Narratives of Language and the Brain, 1825–1926* (Princeton: Princeton University Press, 2000).

Jalland, Pat, *Death in the Victorian Family* (Oxford: Oxford University Press, 1996).

James, Frank A. J. L., *The Common Purposes of Life: Science and Society at the Royal Institution of Great Britain* (Aldershot: Ashgate Publishing, 2002).

Jewson, Nicholas, 'Medical knowledge and the patronage system in eighteenth century England', *Sociology* 8 (1974) 369–85.

——, 'The disappearance of the sick-man from medical cosmology, 1770–1870', *Sociology* 10 (1976) 225–44.

Jordanova, Ludmilla, 'The social construction of medical knowledge' *Journal of the Society of the Social History of Medicine* 8 (1995) 361–82.

——, *Sexual Visions: Images of Gender in Science and Medicine Between the Eighteenth and Twentieth Centuries* (London: Harvester Wheatsheaf, 1989).

Kane, M., Milne, R. and A. F. Smith, *Exploring Professional Boundaries in Anaesthetics*, Draft report 2003, http://www.mbhc.nhs.uk/dept, accessed April 2005.

Kane, M. and A. F. Smith, 'An American tale – professional conflicts in anaesthesia in the United States: Implications for the United Kingdom', *Anaesthesia* 59 (2004) 793–802.

Keys, Thomas E., *The History of Surgical Anaesthesia* (New York: Schumans, 1945).

King, Roger, *The Making of the Dentiste, c.1650–1750*, (Aldershot: Ashgate Publishing, 1998).

Kirkup, J., 'Surgery before general anaesthesia' in Mann 1988.

Knight, David, *Humphry Davy* (Oxford: Blackwell, 1992).

Knight, Paul R. and Douglas R. Bacon, 'An unexplained death, Hannah Greener and Chloroform', *Anaesthesiology* 96 (2002) 1250–3.

Kuhlmann, Ellen, 'The rise of German dental professionalism as a gendered project: How scientific progress and health policy evoked change in gender relations, 1850–1919', *Medical History* 45 (2001) 441–60.

Lane, Joan, *The Making of the English Patient: A Guide to Sources for the History of Medicine* (Stroud: Sutton Publishing Ltd, 2000).
——, *A Social History of Medicine: Health, Healing and Disease in England, 1750–1950* (London: Routledge, 2001).
Lawrence, Christopher, 'The nervous system and society in the Scottish enlightenment' in Barry Barnes and Steve Shapin eds, *Natural Order: Historical Studies of Scientific Culture* (Beverly Hills, London: Sage Publications, 1979).
——, 'Incommunicable knowledge: Science, technology and the clinical art in Britain, 1850–1914', *Journal of Contemporary History* 20 (1985) 503–20.
——, 'The power and the glory: Humphry Davy and romanticism', in Andrew Cunningham and Nicholas Jardine eds, *Romanticism and the Sciences* (Cambridge: Cambridge University Press, 1990).
—— (ed.) *Medical Theory, Surgical Practice: Studies in the History of Surgery* (London and New York: Routledge, 1992).
——, 'Democratic, divine and heroic: the history and historiography of surgery', in Lawrence 1992a.
——, 'Alfred Levy and chloroform death', in Lawrence 1992b.
——, *Medicine in the Making of Modern Britain, 1700–1920* (London: Routledge, 1994).
——, 'Medical minds, surgical bodies: Corporeality and the doctors', in Christopher Lawrence and Steven Shapin eds, *Science Incarnate, Historical Embodiments of Natural Knowledge* (Chicago and London: University of Chicago Press, 1998).
——, 'A tale of two sciences: Bedside and bench in twentieth-century Britain', *Medical History* 43 (1999) 412–49.
Lawrence, Christopher and Ghislaine Lawrence, *No Laughing Matter: Historical Aspects of Anaesthesia* (London: Wellcome Institute for the History of Medicine, 1987).
Lawrence, Christopher and Steven Shapin (eds), *Science Incarnate: Historical Embodiments of Natural Knowledge* (Chicago and London: University of Chicago Press, 1998).
Lawrence, Christopher and Tom Treasure, 'Surgeons' in Roger Cooter and John Pickstone eds, *Companion to Medicine in the Twentieth Century* (London and New York: Routledge, 2003).
Lawrence, Susan C., *Charitable Knowledge* (Cambridge: Cambridge University Press, 1996).
Leake, Chauncey D., 'The historical development of surgical anaesthesia', *The Scientific Monthly* 20 (1925) 304–28.
Leavitt, Judith W., *Brought to Bed: Childbearing in America 1750 to 1950* (Oxford: Oxford University Press, 1986).
Lesch, J. E., *Science and Medicine in France: The Emergence of Experimental Physiology* (Cambridge MA: Harvard University Press, 1984).
Lewis, J. M., 'Early history of ether and chloroform anaesthesia from Welsh newspapers', in Barr *et al.*, 1996.
Lomax, Elizabeth, *History of Children's Hospitals* (London: Wellcome Institute, 1996).
Longford, Elizabeth, *Queen Victoria* (New York: Harper and Row, 1964).
Loudon, Irvine, *Medical Care and the General Practitioner 1750–1850* (Oxford: Oxford University Press, 1986).

——, *Death in Childbirth: An International Study of Maternal Care and Maternal Mortality 1800–1950* (Oxford: Oxford University Press, 1992).

—— (ed.), *Western Medicine: An Illustrated History* (Oxford: Oxford University Press, 1997).

——, 'Doctors and their transport, 1750–1914', *Medical History* 45 (2001) 185–206.

Lufkin, Arthur Ward, *A History of Dentistry* (London: Henry Kimpton, 1938).

Macdonald, M. and T. R. Murphy, *Sleepless Souls: Suicide in Early Modern England* (Oxford: Clarendon, 1990).

Mackenzie, Sir James, *The Future of Medicine* (London: Hoddev & Stoughton, 1919).

Maltby, J. Roger, *Notable Names in Anaesthesia* (London: Royal Society of Medicine Press, 2002).

Mann, Ronald D. (ed.), *The History of the Management of Pain* (Carnforth, Lancs UK: Parthenon Publishing Group, 1988).

Manuel, Diana E., *Marshall Hall (1790–1857): Science and Medicine in Early Victorian Society* (Amsterdam-Atlanta: Rodopi, 1996).

——, 'Robert Lee, the uterine nervous system and a wrangle at the Royal Society, 1839–1849', *Journal of the Royal Society of Medicine* 94 (2001) 645–7.

Marland, Hilary, *Medicine and Society in Wakefield and Huddersfield 1780–1870* (Cambridge: Cambridge University Press, 1987).

Matthews, J. Rosser, *Quantification and the Quest for Medical Certainty* (Princeton NJ: Princeton University Press, 1995).

Maulitz, Russell, *Morbid Appearances: The Anatomy of Pathology in the Early Nineteenth Century* (Cambridge: Cambridge University Press, 1987).

McCullough, Laurence B., 'John Gregory's Medical Ethics', in Baker, Porter and Porter, 1993.

Medvei, Victor Cornelius and John L. Thornton (eds), *The Royal Hospital of Saint Bartholomew 1123–1973* (London: The Royal Hospital of St Bartholomew, 1974).

Merrington, W. R., *University College Hospital and its Medical School: A History* (London: Heinemann, 1976).

Miller, Jonathan, 'Going unconscious' in Robert B Silvers, ed. *Hidden Histories of Science* (New York: New York Review, 1995).

Morantz-Sanchez, Regina, 'Making it in a man's world: The late-nineteenth century surgical career of Mary Amanda Dixon Jones', *Bulletin of the History of Medicine* 69 (1995) 542–68.

Morrell, Jack, Arnold Thackray, *Gentlemen of Science: Early Years of the British Association for the Advancement of Science* (Oxford: Oxford University Press, 1981).

Morris, David B., *The Culture of Pain* (Berkeley and Los Angeles: University of California Press, 1991).

Morris, R. J., *Cholera 1832* (London: Croom Helm, 1976).

Mort, Frank, *Dangerous Sexualities: Medico-Moral Politics in England since 1830* (London and New York: Routledge, 2000).

Moscucci, Ornella, *The Science of Woman: Gynaecology and Gender in England, 1800–1929* (Cambridge: Cambridge University Press, 1990).

Nettleton, Sarah, 'Protecting a vulnerable margin: Towards an analysis of how the mouth came to be separated from the body', *Sociology of Health & Illness* 10 (1988) 156–69.

Newman, Charles, *The Evolution of Medical Education in the Nineteenth Century* (Oxford: Oxford University Press, 1957).
Nicolson, Malcolm, 'The introduction of percussion and stethoscopy to early nineteenth-century Edinburgh', in Bynum and Porter 1993a.
Ogburn, William F. and Dorothy Thomas, 'Are inventions inevitable? A note on social evolution', *Political Science Quarterly* 37 (1922) 83–98.
Oppenheim, Janet, *Shattered Nerves: Doctors, Patients and Depression in Victorian England* (New York, Oxford: Oxford University Press, 1991).
Owen, Buddug, 'From Wrexham to Rheims in 1847', *Proceedings of the History of Anaesthesia Society* 15 (1994) 37–40.
Papper, E. M., *Romance, Poetry and Surgical Sleep* (Westport, Connecticut, 1995).
Parry, Jose, Noel Parry, *The Rise of the Medical Profession, A Study of Collective Mobility* (London: Croom Helm, 1976).
Parssinen, Terry M., *Secret Passions, Secret Remedies: Narcotic Drugs in British Society, 1820–1930* (Manchester: Manchester University Press, 1983).
Paul, Harry W., *Bacchic Medicine* (Amsterdam-New York: Rodopi, 2001).
Pernick, Martin S., *A Calculus of Suffering: Pain, Professionalism and Anesthesia in Nineteenth Century America* (New York: Columbia University Press, 1985).
Peterson, M. Jeanne, *The Medical Profession in Mid-Victorian England*, (California: University of California Press, 1978).
Pickstone, J. V., 'Bureaucracy, liberalism and the body in post-revolutionary France: Bichat's physiology and the Paris school of medicine', *History of Science* 19 (1981) 115–42.
—— (ed.), *Medical Innovations in Historical Perspective* (New York, Basingstoke: Macmillan, 1992).
——, 'The biographical and the analytical: Towards a historical model of science and practice in modern medicine', in I. Lowy ed. *Medicine and Change: Historical and Sociological Studies of Medical Innnovation* (Paris: Inserm, 1993), 23–46.
——, 'Ways of knowing: Towards a historical sociology of science, technology and medicine', *British Journal of the History of Science* 26 (1993) 433–58.
——, *Ways of Knowing: A New History of Science, Technology and Medicine* (Manchester: Manchester University Press, 2000).
——, 'Science in nineteenth-century England: Plural configurations and singular politics', in Martin Daunton, ed. *The Organisation of Knowledge in Victorian Britain* (Oxford: Oxford University Press, 2005).
Pinker, Robert, *English Hospital Statistics, 1861–1938* (London: Heinemann, 1966).
Polanyi, Michael, *Personal Knowledge* (Chicago: University of Chicago Press, 1974).
Poovey, Mary, *Making a Social Body, British Cultural Formation 1830–64* (Chicago and London: University of Chicago Press, 1995).
Poovey, Mary, *Uneven Developments: The Ideological Work of Gender in Mid-Victorian England* (London: Virago, 1989).
Poovey, Mary, ' "Scenes of an Indelicate Character": The medical "treatment" of Victorian women', in Catherine Gallagher and Thomas Lacqueur eds, *The Making of the Modern Body* (Berkeley and Los Angeles: University of California Press, 1987).
Porter, Roy, *English Society in the Eighteenth Century* (London: Allen Lane, 1982).
——, 'The patient's view: Doing medical history from below', *Theory and Society*, 14 (1985) 175–98.

—— (ed.), *Patients and Practitioners: Lay Perceptions of Medicine in Pre-Industrial Society* (Cambridge: Cambridge University Press, 1985).
——, *Disease, Medicine and Society in England 1550–1860* (London: Allen Lane, 1987).
——, *The Enlightenment* (Hampshire: Macmillan, 1990).
——, *Doctor of Society*, (London and New York: Routledge, 1992).
——, 'The rise of the physical examination', in Bynum and Porter 1993a.
——, *London: A Social History* (Harmondsworth: Penguin Books, 1996).
——, *The Greatest Benefit to Mankind: A Medical History of Humanity from Antiquity to the Present* (London: Fontana Press, 1999).
——, *Flesh in the Age of Reason* (London: Allen Lane, 2003).
Porter, Roy and Dorothy Porter, *Patient's Progress* (Stanford: Stanford University Press, 1989).
Porter, Roy and Mikulas Teich (eds), *Drugs and Narcotics in History* (Cambridge: Cambridge University Press, 1995).
Prest, John, 'Peel, Sir Robert, second baronet (1788–1850)', *Oxford Dictionary of National Biography* (Oxford: Oxford University Press, 2004).
Prys-Roberts, Cedric and Burnell R. Brown Jr (eds), *International Practice of Anaesthesia*, 2 vols (Oxford: Butterworth Heinemann, 1996).
Reiser, Stanley J., *Medicine and the Reign of Technology* (Cambridge: Cambridge University Press, 1978).
Rey, Roselyne, *The History of Pain*, trans. Louise Elliott Wallace, J. A. Cadden and S. W. Cadden (Cambridge, Mass.: Harvard University Press, 1995).
Richards, Stewart, 'Anaesthetics, ethics and aesthetics: Vivisection in the late nineteenth-century British laboratory', in Cunningham and Williams 1992.
Richardson, Ruth, *Death, Dissection and the Destitute: A Political History of the Human Corpse* (London: Routledge & Kegan Paul, 1987).
Risse, Guenter B., *Hospital Life in Enlightenment Scotland: Care and Teaching at the Royal Infirmary* (Cambridge: Cambridge University Press, 1986).
Risse, Guenter B. and John H. Warner, 'Reconstructing clinical activities: Patient records in medical history', *Journal of the Social History of Medicine* 5 (1992) 183–205.
Rosen, George, *The Specialization of Medicine with Particular Reference to Ophthalmology* (New York: Arno Press, 1972).
Rosenberg, Charles, 'The therapeutic revolution: Medicine, meaning and social change in 19th century America', in Vogel and Rosenberg 1979.
Rudwick, Martin, *The Great Devonian Controversy: The Shaping of Scientific Knowledge Among Gentlemanly Specialists* (Chicago: University of Chicago Press, 1985).
Rupke Nicolaas A. (ed.), *Vivisection in Historical Perspective* (London: Croom Helm, 1987).
Ruse, Michael, 'Darwin's debt to philosophy: An examination of the influence of the philosophical ideas of J. F. W. Herschel and W. Whewell on the development of Charles Darwin's theory of evolution', *Studies in the History and Philosophy of Science*, 6 (1975) 159–81.
Rushman, G. B. N. J. H. Davis, R. S. Atkinson, *A Short History of Anaesthesia: The First 150 Years*, (Oxford, Boston: Butterworth Heinemann, 1996).
Schaffer, Simon, 'Scientific discoveries and the end of natural philosophy', *Social Studies of Science* 16 (1986) 387–420.
Schlich, Thomas, *Surgery, Science and Industry: A Revolution in Fracture Care, 1950s–1990s* (Basingstoke: Palgrave Macmillan, 2002).

Schlich, Thomas and Ulrich Troêhler (eds), *The Risks of Medical Innovation: Risk Perception and Assessment in Historical Context* (London and New York: Routledge, 2006).
Schofield, Robert E., *Mechanism and Materialism* (Princeton, NJ: Princeton University Press, 1970).
Scott, W., M. Vickers, H. Draper (eds), *Ethical Issues in Anaesthesia* (Oxford: Butterworth, 1994).
Secher, O., '46 "first" anaesthetics in the world', *Acta Anaesthesiologica Scandinavica* 34 (1990) 552–6.
Shepherd, J. A., *Simpson and Syme of Edinburgh* (Edinburgh: E. and S. Livingstone, 1969).
——, *The Crimean Doctors: A History of the British Medical Services in the Crimean War* (Liverpool: Liverpool University Press, 1991).
Showalter, Elaine, *The Female Malady: Women, Madness and English Culture, 1830–1980* (Virago, London, 1987).
Shortt, S. E. D., 'The professionalization of Anglo-American medicine in the nineteenth century', *Medical History* 27 (1983) 51–68.
Shrycock, Richard H., 'The history of quantification in medical science', *Isis* 52 (1961) 215–37.
Silvers, Robert B (ed.), *Hidden Histories of Science* (New York: New York Review, 1995).
Simpson, Myrtle, *Simpson the Obstetrician: A Biography* (London: Gollancz, 1972).
Smith, W. D. A., *Under the Influence* (London: Macmillan, 1982).
Snow, Stephanie J., *John Snow MD, 1813–1858, The Emergence of the Medical Profession*, PhD diss. University of Keele, 1995.
——, John Snow MD 1813–1858, Part I 'A Yorkshire childhood and family life', *Journal of Medical Biography* 8 (2000) 27–31.
——, John Snow MD 1813–1858, Part II 'Becoming a doctor – his medical training a early years of practice', *Journal of Medical Biography* 8 (2000) 71–77.
Spillane, John D., *The Doctrine of the Nerves* (Oxford: Oxford University Press, 1981).
St George, Andrew, *The Descent of Manners: Etiquette, Rules and the Victorians* (London: Chatto and Windus, 1993).
Stanley, Peter, *For Fear of Pain: British Surgery 1790–1850* (Amsterdam-New York, 2003).
Stansfield, Dorothy, *Thomas Beddoes MD 1760–1808: Chemist, Physician, Democrat* (Dordrecht: Reidel, 1984).
Stevens, Rosemary, *Medical Practice in Modern England: The Impact of Specialization and State Medicine* (New Haven and London: Yale University Press, 1966).
——, 'The changing idea of a medical specialty', *Transactions and Studies of the College of Physicians of Philadelphia* V (1980) 159–77.
Stevenson, Lloyd G., 'Suspended animation and the history of anaesthesia', *Bulletin of the History of Medicine* 49 (1979) 482–511.
Steward, K. A., 'The anaesthetist's table', in Barr *et al.*, 1996.
Sykes, W. S., *Essays on the First Hundred Years of Anaesthesia*, 3 vols (Edinburgh: Livingstone, 1960, 1961, 1982).
Symons, John, 'The quest for Hickman', *Medical History* 43 (1999) 95–107.
Thatcher, Virginia, *History of Anaesthesia with Special Emphasis on the Nurse Specialist* (Philadelphia: Lippincott, 1953).

Thomas, K. Bryn, 'The clover/snow collection', *Anaesthesia* 27 (1972) 436–49.
——, *The Development of Anaesthetic Apparatus* (Oxford: Association of Anaesthetists, 1975).
Thornton, J. L., 'Development of department of anaesthetics', *St Bartholomew's Hospital Journal* (1959 January) 13–59.
Trombley, Stephen, *Sir Frederick Treves: The Extra-Ordinary Edwardian* (London: Routledge, 1989).
Trohler, Ulrich 'Surgery (Modern)' in Bynum and Porter, vol. 2 (1993) 984–1028.
——, *To Improve the Evidence of Medicine: The 18th Century British Origins of a Critical Approach* (Edinburgh: Royal Society of Physicians of Edinburgh, 2000).
Trombley, Stephen, *Sir Frederick Treves: The Extra-Ordinary Edwardian* (London: Routledge, 1989).
Uglow, Jenny, *The Lunar Men: The Friends Who Made the Future* (London: Faber and Faber, 2002).
Van Zwanenberg, David, 'The training and careers of those apprenticed to apothecaries in Suffolk, 1815–1858', *Medical History* 27 (1983) 139–50.
Vincent, David, *Bread, Knowledge and Freedom: A Study of Nineteenth-Century Working Class Autobiography* (London: Methuen, 1981).
——, *The Culture of Secrecy: Britain 1832–1998*, (Oxford: Oxford University Press, 1998).
Vinten-Johansen, 'Peter Howard Body, Nigel Paneth' Stephen Rachman, Michael Rip, *Cholera, Chloroform and the Science of Medicine: A Life of John Snow* (Oxford: Oxford University Press, 2003).
Vogel, Morris and Charles Rosenberg (eds), *The Therapeutic Revolution* (New York: University of Pennsylvania Press, 1979).
Waddington, Ivan, 'The development of medical ethics: A sociological analysis', *Medical History* 19 (1975) 36–51.
——, *The Medical Profession in the Industrial Revolution* (Dublin: Gill and Macmillan, 1984).
Walker, J. B. R., 'Etherization as seen by the local press', *Proceedings of the History of Anaesthesia Society* 19 (1996) 25–9.
Warner, John Harley, 'Therapeutic explanation and the Edinburgh bloodletting controversy: Two perspectives on the medical meaning of science in the mid-nineteenth century', *Medical History* 24 (1980) 241–58.
——, *The Therapeutic Perspective. Medical Practice, Knowledge and Identity in America, 1820–1885* (Cambridge, Mass.: Harvard University Press, 1986).
——, 'The idea of science in English medicine: The "decline of science" and the rhetoric of reform, 1815–1845', in French and Wear 1992.
——, 'Review of the Case-Books', *Medical History*, 40 (1996), 382–3.
——, 'Orthodoxy and otherness: Homeopathy and regular medicine in nineteenth-century America', in Robert Jutte, Guenter B. Risse, John Woodward eds, *Culture, Knowledge and Healing, Historical Perspectives of Homeopathic Medicine in Europe and North America* (Sheffield: European Association for the History of Medicine and Health Publications, 1998).
Waserman, Manfred, 'Sir James Y Simpson and London's "conservative and so curiously prejudiced" Dr Ramsbotham', *BMJ* 2 (1980) 158–61.
Wear, Andrew (ed.) *Medicine in Society: Historical Essays* (Cambridge: Cambridge University Press, 1992).
Weatherall, Miles, 'Drug therapies', in Bynum and Porter, vol. 2 (1993) 915–38.

Weinberger, Bernhard Wolf, *An Introduction to the History of Dentistry in America*, 2 vols, (St Louis: C. V. Mosby, 1948).

Weisz, George, 'Mapping medical specialization in Paris in the nineteenth and twentieth centuries', *Journal of the Social History of Medicine* 7 (1994) 177–212.

——, 'The emergence of medical specialisation in the nineteenth century', *Bulletin of the History of Medicine* 77 (2003) 536–75.

White, Harvey, 'Surgery in the eighteenth and nineteenth centuries', in Victor C. Medvei and John L. Thornton eds, *The Royal Hospital of Saint Bartholomew* (London: Libraries of Saint Bartholomew's Hospital Medical Colleges, 1974), 205–21.

Wiener, Martin J., *Reconstructing the Criminal: Culture, Law and Policy in England, 1830–1914* (Cambridge: Cambridge University Press, 1994).

Wilkinson, D. J., 'A strange little book', *Anaesthesia* 58 (2003) 36–41.

Williams, Guy, *The Age of Miracles* (London: Academy Chi. Pubs, 1987).

Wilson, Gwen, *One Grand Chain: The History of Anaesthesia in Australia, 1846–1962* (Melbourne: Australian and New Zealand College of Anaesthetists, 1995).

Wilson, John B, 'A surgeon's private practice in the nineteenth century', *Medical History*, 31 (1987) 349–53.

Winter, Alison, 'Ethereal epidemic: Mesmerism and the introduction of inhalation anaesthesia to early Victorian London', *Journal of the Social History of Medicine* 4 (1991) 1–28.

——, *Mesmerized: Powers of Mind in Victorian Britain* (Chicago: University of Chicago Press, 1998).

Witten, Renate, 'The origins of homeopathy in German', in W. F. Bynum and V. Nutton (eds), *Essays in the History of Therapeutics* (Rodopi Amsterdam-Atlanta, GA 1991).

Witz, Anne, *Professions and Patriarchy* (London: Routledge, 1992).

Wohl, Anthony S., *Endangered Lives: Public Health in Victorian Britain* (London: J. M. Dent, 1983).

Wolfe, Richard J., *Tarnished Idol, William Thomas Green Morton and the Introduction of Surgical Anesthesia: A Chronicle of the Ether Controversy* (San Anselmo, California: Norman Publishing, 2001).

Woods, Robert and John Woodward (eds), *Urban Disease and Mortality in Nineteenth Century England* (London, New York: Batsford Academic and Educational Ltd, 1984).

Woodward, J. H., *To do the Sick No Harm: A Study of the British Voluntary Hospital System to 1875* (London: Routledge and Kegan Paul, 1974).

Woollam, C. H. M., 'The sister anaesthetists of Norwich', *Anaesthesia* 57 (2002) 984–94.

Woolcock, Helen. R M. John Thearle and Kay Saunders, '"My beloved chloroform". Attitudes to childbearing in Colonial Queensland: A case study', *Journal of the Society of the Social History of Medicine* 10 (1997) 437–56.

Worboys, Michael *Spreading Germs: Disease Theories and Medical Practice in Britain, 1865–1900* (Cambridge: Cambridge University Press, 2000).

——, 'British medicine and its past at Queen Victoria's jubilees and the 1900 centennial', *Medical History* 45 (2001) 461–82.

Wright, Peter, 'The radical sociology of medicine', *Social Studies of Science* 109 (1980) 103–20.

Wright, Peter and Andrew Treacher (eds), *The Problem of Medical Knowledge: Examining the Social Construction of Medicine* (Edinburgh: Edinburgh University Press, 1982).

Yalton, Y. W., *Thinking Matter: Materialism in 18th century Britain* (Oxford: Basil Blackwell, 1984).

Yeo, Geoffrey, *Images of Bart's* (Historical Publications Ltd in association with the Archives Department, St Bartholomew's Hospital, 1992).

Young, Hugh H., 'Crawford W. Long: The pioneer in ether anaesthesia', *Bulletin of the History of Medicine* 12 (1942) 191–225.

Young, T. M., *A Short History of the Section of Anaesthetics of the Royal Society of Medicine and the Society of Anaesthetists* (London: Royal Society of Medicine, 1996).

Youngson, A. J., *The Scientific Revolution in Victorian Medicine* (London: Croom Helm, 1979).

Zuck, D., 'Clover, Joseph Thomas (1825–1882)', *Dictionary of National Biography*, (Oxford: Oxford University Press, 2004).

# Index